Springer Theses

Recognizing Outstanding Ph.D. Research

Aims and Scope

The series "Springer Theses" brings together a selection of the very best Ph.D. theses from around the world and across the physical sciences. Nominated and endorsed by two recognized specialists, each published volume has been selected for its scientific excellence and the high impact of its contents for the pertinent field of research. For greater accessibility to non-specialists, the published versions include an extended introduction, as well as a foreword by the student's supervisor explaining the special relevance of the work for the field. As a whole, the series will provide a valuable resource both for newcomers to the research fields described, and for other scientists seeking detailed background information on special questions. Finally, it provides an accredited documentation of the valuable contributions made by today's younger generation of scientists.

Theses are accepted into the series by invited nomination only and must fulfill all of the following criteria

- They must be written in good English.
- The topic should fall within the confines of Chemistry, Physics, Earth Sciences, Engineering and related interdisciplinary fields such as Materials, Nanoscience, Chemical Engineering, Complex Systems and Biophysics.
- The work reported in the thesis must represent a significant scientific advance.
- If the thesis includes previously published material, permission to reproduce this must be gained from the respective copyright holder.
- They must have been examined and passed during the 12 months prior to nomination.
- Each thesis should include a foreword by the supervisor outlining the significance of its content.
- The theses should have a clearly defined structure including an introduction accessible to scientists not expert in that particular field.

More information about this series at http://www.springer.com/series/8790

Nicola Di Iorio

New Organocatalytic Strategies for the Selective Synthesis of Centrally and Axially Chiral Molecules

Doctoral Thesis accepted by
the University of Bologna, Italy

Springer

Author
Dr. Nicola Di Iorio
Department of Industrial Chemistry
"Toso Montanari"
University of Bologna
Bologna
Italy

Supervisor
Prof. Dr. Paolo Righi
Department of Industrial Chemistry
"Toso Montanari"
University of Bologna
Bologna
Italy

ISSN 2190-5053 ISSN 2190-5061 (electronic)
Springer Theses
ISBN 978-3-030-09103-3 ISBN 978-3-319-74914-3 (eBook)
https://doi.org/10.1007/978-3-319-74914-3

Printed on acid-free paper

This Springer imprint is published by the registered company Springer International Publishing AG part of Springer Nature
The registered company address is: Gewerbestrasse 11, 6330 Cham, Switzerland

Find a synthesis, that's my objective
But conformers aren't fully elective:
Only one is desired.
A reaction's required
That's atropo-enantioselective

Supervisor's Foreword

Asymmetric catalysis is a very important branch of chemistry. It allows for greener and more sustainable processes and it is usually divided into three major areas that are enzymatic, metal, and organic catalysis. The following Ph.D. thesis concerns the use of asymmetric organic catalysis for the direct, enantioselective synthesis of complex chiral molecules. Organocatalysis is quite a modern and "hot" topic in the chemical community, it is in constant expansion and its use has been extended to very interesting areas like vinylogous reactivity and atropisomerism. Vinylogous systems are very important for their synthetic applications but pose a number of challenges, the most notable of them are their reduced reactivity and the reduced stereocontrol at this kind of positions. On the other hand, atropisomeric systems are maybe even more important because of the huge potential they have as drugs, ligands, and catalysts. Chemists have only recently "recognized" the importance of these two areas and are putting a big deal of effort in order to study them and the challenges they pose. For this reason, the work presented in this thesis is very important, as it addresses the many aspects of both vinylogy and atropisomerism. Specifically, the vinylogous Michael addition of nonsymmetric 3-alkylidene oxindoles to nitroolefines was optimized and its mechanism was studied and explained in detail. Next, the desymmetrization of hindered *N*-(2-*t*-butylphenyl) maleimides is described. By using many nucleophiles and activation modes, it is shown how general and effective this strategy is to obtain non-biarylic, unconventional, axially chiral molecules. Finally, it reports the first direct, atroposelective synthesis of $C(sp^2)$-$C(sp^3)$ atropisomers. These compounds have only been studied under a conformational point of view and they were only synthetized as racemic mixtures under very drastic conditions. This thesis describes the Friedel–Crafts reaction of hindered indenones and β-naphthols that turned out to be a mild and general method for the selective synthesis of this kind of atropisomers. Thet thesis also features an extensive introduction on the general aspects of chirality and organocatalysis and an equally extensive experimental section in order to better guide nonexperts in the understanding of the discussion part and to allow reproducibility of experiments.

Bologna, Italy
April 2017

Prof. Dr. Paolo Righi

Preface

All the work discussed in this thesis was done for two possible purposes: either to develop a new reactivity or to achieve a specific structural feature in the product molecule (or both at the same time). However, regardless of the purpose we had for every project, the tool we employed to reach our target was always asymmetric organocatalysis. Although organocatalysis is an old branch of chemistry, its role in asymmetric synthesis has been rightfully "recognized" by the scientific community just recently (2000), therefore it has experienced a deep and ongoing development only in the last 17 years. Nevertheless, this relatively modern discipline has proven to be a powerful tool for synthetic chemists and nowadays holds its ground brilliantly to the older metal and metallorganic catalysis. Indeed, there has been a huge number of publications per year (>1000/year according to SciFinder) reporting an ever-growing number of activation modes, many of which have been used for the reactions discussed in this elaborately. Consequently, giving a full description of every activation mode would be very lengthy and it is not the purpose of this thesis, on the other hand leaving the reader completely without a background would be inappropriate. Therefore, in order to get straight to the discussion part, the general introduction will be as short as possible with two main sections: the first one concerning chirality (it will be discussed in view of the synthetic issues arising from it, not only from a generic point of view) with special attention to noncentral chirality and atropisomers, followed by the second one concerning organocatalysis made of a list of the principal activation modes each coupled with a single relevant example for better understanding. For a more detailed insight on a general topic, the reader will have to rely on the references that will be given at the end of the chapter. The single projects will not be discussed in a chronological order, instead they will be listed following a topic/purpose order, so after the first project entirely about developing an organocatalytic vinylogous reactivity, a second one will be discussed concerning both organocatalytic and structural aspects and then a final one whose main focus and novelty lies in the peculiar structure of the products. As mentioned above, some of the work in this thesis concerns vinylogy. The concept of vinylogy cannot be considered as a specific activation mode itself but as an extension of it, therefore it will not be treated in the introduction and will be dealt with later in the discussion part. Finally, from a geometrical point of view, a center, an axis, or a plane cannot be chiral. The correct way of naming an element that makes a molecule chiral would be “stereogenic”, but there may be occasions where such elements will be called “chiral” or “asymmetric” to avoid redundant repetitions of the word “stereogenic”.

Bologna, Italy Nicola Di Iorio

Parts of this thesis have been published in the following articles:

- N. Di Iorio, G. Filippini, A. Mazzanti, P. Righi, G. Bencivenni: "Controlling the C(sp^3)-C(sp^2) axial conformation in the enantioselective Friedel-Crafts-type alkylation of β-Naphthols with inden-1-ones" Org. Lett., 2017, 19 (24), pp 6692–6695
- N. Di Iorio, L. Soprani, S. Crotti, E. Marotta, A. Mazzanti, P. Righi, G. Bencivenni: "Michael addition of oxindoles to *N*-(2-*tert*-butylphenyl)maleimides: efficient desymmetrization for the synthesis of atropisomeric succinimides with quaternary and tertiary stereocenters" *Synthesis* (invited paper), **2017**, 49 (07), 1519–1530
- N. Di Iorio, F. Champavert, A. Erice, P. Righi, A. Mazzanti, G. Bencivenni: "Targeting remote axial chirality control of *N*-(2-*tert*-butylphenyl)succinimides by means of Michael addition type reaction" *Tetrahedron*, **2016**, 72, 5191–5201
- N. Di Iorio, P. Righi, S. Ranieri, A. Mazzanti, R. G. Margutta, G. Bencivenni: "Vinylogous reactivity of oxindoles bearing nonsymmetric alkylidene groups" *Journal of Organic Chemistry*, **2015**, 80 (14), 7158–7171
- N. Di Iorio, P. Righi, A. Mazzanti, M. Mancinelli, A. Ciogli, G. Bencivenni: "Remote control of axial chirality: Aminocatalytic desymmetrization of *N*-arylmaleimides via vinylogous Michael addition" *Journal of American Chemical Society*, **2014**, 136 (29), 10250–10253

Acknowledgements

My first thanks go to those who have, in whichever way, scientifically and directly contributed to the development and making of this thesis and, more generally, to the experimental work behind it. Be it for giving me a fume hood, for helping me with complicated NMR experiments, for sharing the practical work, for teaching me something, for giving me precious advices, or simply for their availability, I wanted to thank Paolo, Giorgio, Andrea, my fellow Ph.D. colleagues, and all the students.

My last and most important thanks go to those who gave me something more than practical or theoretical science, to those who have shared their minds, joys, sweat, pain, and tears with me, to those that made me grow better, and to those that grew better with me. In other words, to everyone that somehow made my Ph.D. journey a worthy, enjoyable, and meaningful one. In regard to this, out of all people, I wanted to specifically mention Carla, Sarona, and Boa for the patience, the love, the cheerfulness, the friendship, and the growth that we have shared and continue to share now.

Contents

Abbreviations

Ac	Acetyl
ACDC	Asymmetric counterion-directed catalysis
AcO	Acetate
AcOH	Acetic acid
AIBN	Azaisobutironitrile
Ar	Aryl
BINOL	1,1'-bi-2-naphthol
Boc	*tert*-Butyloxycarbonyl
CA	Cinchonine
Calc	Calculated, calculation
CBz	Carboxybenzoil
CDA	Cinchonidine
DA	Diels–Alder
DCM	Dichloromethane
DHCA	Dihydrocinchonine
DHCDA	Dihydrocinchonidine
DHQA	Dihydroquinine
DHDQA	Dihydroquinidine
DIAD	Diisopropyl azodicarboxylate
DKR	Dynamic kinetic resolution
DMAP	*N,N,*Dimethylamino pyridine
DMPU	*N,N*-Dimethylpropylene urea
DMSO	Dimethyl sulfoxide
d.r.	Diastereomeric ratio
E	Electrophile
ee	Enantiomeric excess
ent	Enantiomer
er	Enantiomeric ratio
Et	Ethyl
EtOH	Ethanol

EWG	Electron-withdrawing group
Exp	experiment, experimental
EXSY	Exchange spectroscopy
gCOSY	Gradient correlation spectroscopy
GS	Ground state
h	hour
HFIP	Hexafluoroisopropanol
HOMO	Highest occupied molecular orbital
HPLC	High-performance liquid chromatography
HRMS	High-resolution mass spectroscopy
KIE	Kinetic isotope effect
KR	Kinetic resolution
LA	Lewis acid
LUMO	Lowest unoccupied molecular orbital
M	Molar (concentration)
m	*meta*
MBH	Morita–Baylis–Hillman
Me	Methyl
MeOH	Methanol
Moc	Methyloxycarbonyl
MS	Molecular sieves
MTBE	Methyl, *tert*-butyl ether
NBS	*N*-bromo succinimide
NHC	*N*-heterocyclic carbene
NMR	Nuclear magnetic resonance
NOESY	Nuclear Overhauser effect spectroscopy
NPA	*N*-phosphoramide
Nu	Nucleophile
o	*ortho*
p	*para*
PA	Phosphoric acid
Ph	Phenyl
PhBr	Bromobenzene
PhCl	Chlorobenzene
PhF	Fluorobenzene
PTC	Phase-transfer catalysis, Phase-transfer catalyst
QA	Quinine
QDA	Quinidine
RDS	Rate determining step
r.t.	Room temperature
SET	Single electron transfer
SPA	Spirophosphoric acid
SQ	Squaramide
t	*tert*
TBDMS	*tert*-butyldimethylsilyl

TEA	Triethylamine
Tf	Triflate
TFA	Trifluoroacetic acid
TFAA	Trifluoroacetic anhydride
THF	Tetrahydrofuran
TLC	Thin layer chromatography
TMS	Trimethylsilyl, tetramethylsilane
TS	Transition state
TU	Thiourea
XRD	X-ray diffraction

Chapter 1
Introduction

I would like to start this elaborate by posing the most fundamental question: why do we study chirality? The answer is because Nature is chiral and, as part of it ourselves, it is imperative to understand it and its chiral mechanisms. Although chirality can generally be observed on a macroscale (e.g. conch shells, human ears, human hands…), our interest as chemists lies in its microscopic side: chiral molecules and how to selectively and efficiently synthesize them.

1.1 Chirality and its Forms: Central, Axial and Unconventional[1]

Chirality is the property of an object of being non-superimposable on its mirror image [2]. A chiral object can then exist as two enantiomers which correspond to the mentioned mirror images. On a molecular scale, there are many ways to meet this definition and for a molecule to be chiral it must possess a stereogenic element, which also defines the type of chirality the molecule shows, like an atom, an axis, a helix or a plane.

1.1.1 Central Chirality

It manifests in a molecule bearing an atom with four different substituents (i.e. a stereogenic atom), it is the most widespread type of chirality and consequently also the easiest to visualize and understand (Fig. 1.1).

[1] For further reading on chirality see: Eliel and Wilen [1].

N. Di Iorio, *New Organocatalytic Strategies for the Selective Synthesis of Centrally and Axially Chiral Molecules*, Springer Theses,
https://doi.org/10.1007/978-3-319-74914-3_1

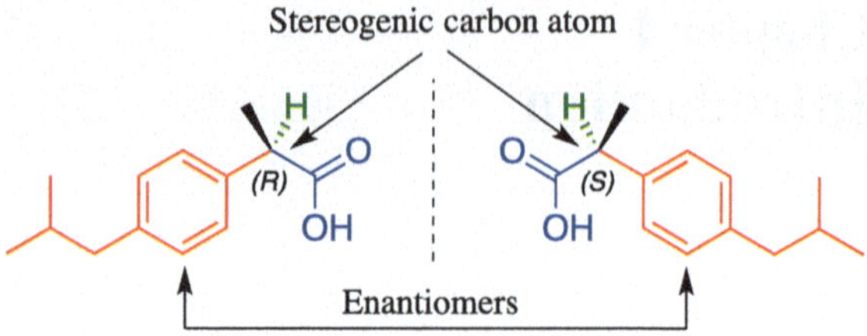

Fig. 1.1 The two enantiomers of Ibuprofen arising from the highlighted asymmetric carbon

Being tetravalent and very abundant, Carbon (in its tetrahedral form) is the most common stereogenic atom to be found, and all the products showed in this thesis possess central chirality on Carbon atoms, but there are many reported cases of molecules possessing chiral heteroatoms (e.g. Nitrogen or Phosphorus)[2].

What makes central chirality unique is the possibility of easily having multiple stereocenters in the same structure. This is possible in theory also for the other stereogenic elements introduced earlier, but it is very rare to find molecules (outside of sheer academic exercises) with more than two chiral axes and, even rarer, more than two chiral helixes or planes because to generate them, very hindered and peculiar structures are required. On the contrary, a carbon atom only needs to have four different substituents to be chiral. For synthetic purposes the consequences of this are huge, in fact the selectivity issues arising from the synthesis of a relatively small (compared to other biomolecules) and ubiquitous molecule like cholesterol (Fig. 1.2) are considerable because 256 stereoisomers of it can be obtained.

This may not be an issue for the frontier researcher busy to find new activation modes on very simple molecules, but for the applied synthetic chemist, dealing with and controlling central chirality of ever-growing (in both dimension and complexity) compounds, often proves to be a tough task due to the selectivity issues just mentioned. A remarkable example of such selectivity was published by Enders and coworkers in 2006 [4]. They reported the synthesis of chiral cyclohexenes bearing four consecutive stereocenters (Reaction 1.1) via a one-pot triple cascade reaction with almost complete control over central chirality.

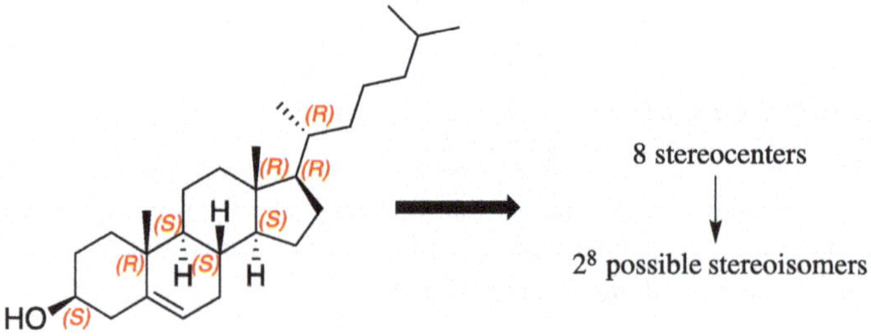

Fig. 1.2 Cholesterol as it occurs in nature and the number of synthetically available stereoisomers

---

[2]For further reading on heteroatoms chirality see: Wolf [3].

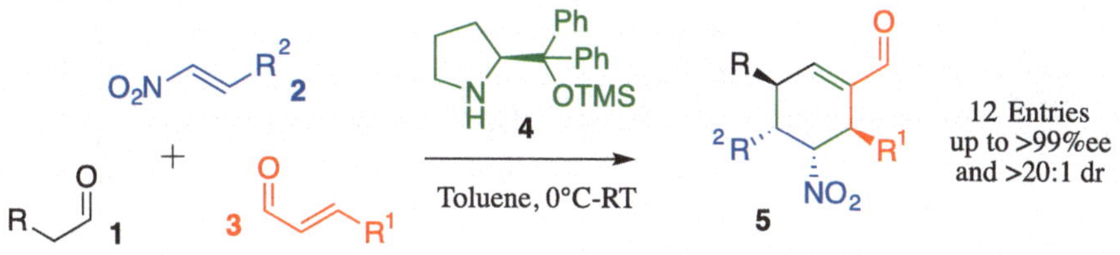

Reaction 1.1 The triple cascade reaction by Enders et al. A nearly complete control over central chirality is achieved

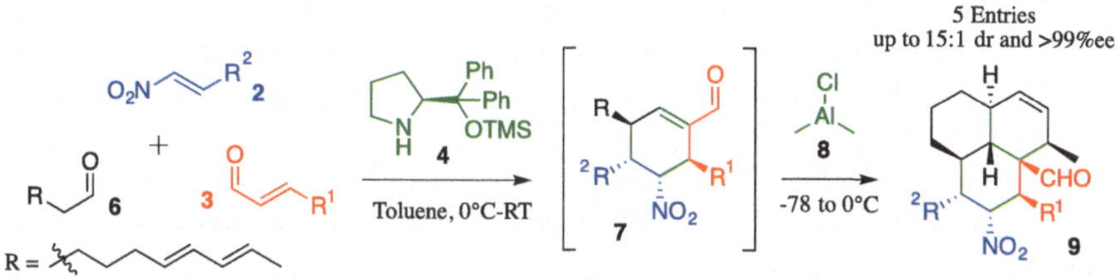

Reaction 1.2 The control of eight stereocenters in the synthesis of tricyclic carbon scaffold

The same group improved this result in 2007 [5] and reported the one pot synthesis of tricyclic carbon scaffolds (Reaction 1.2) featuring a total of eight stereocenters (seven adjacent ones) again with almost complete control over central chirality.[3]

1.1.2 *Axial Chirality*

This less known type of chirality arises from a restricted rotation around a single bond and the molecules possessing such a feature are called "atropisomers" (Fig. 1.3) [2].

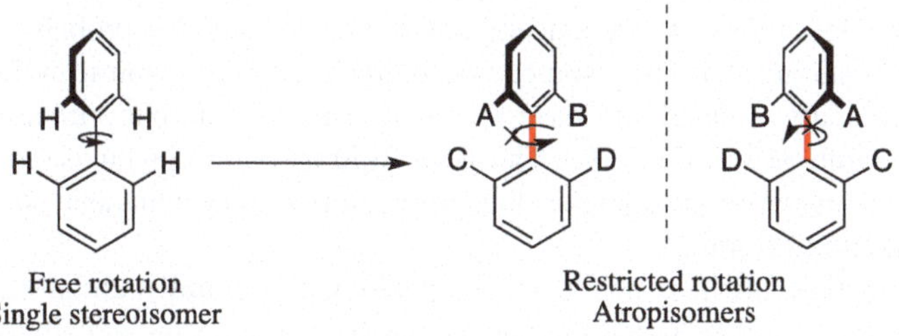

Fig. 1.3 An example of an atropisomeric compound

[3]For more noteworthy examples of controlled central chirality see; (a) Enders et al. [6]; (b) Cassani et al. [7]; (c) Zhou et al. [8]; (d) Chauhan et al. [9]; (e) Reyes et al. [10].

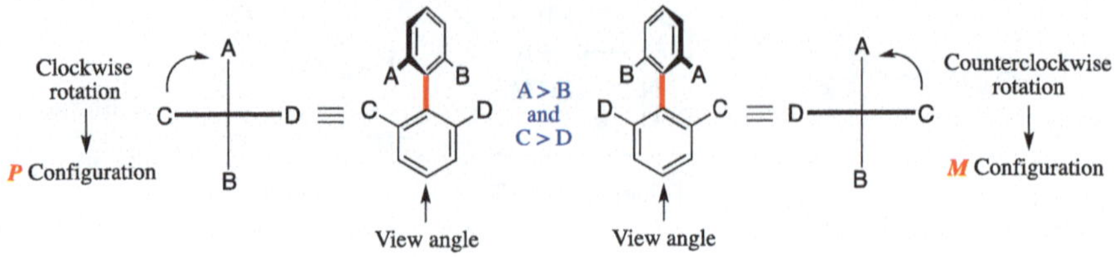

Fig. 1.4 *P*, *M* nomenclature of stereogenic axes

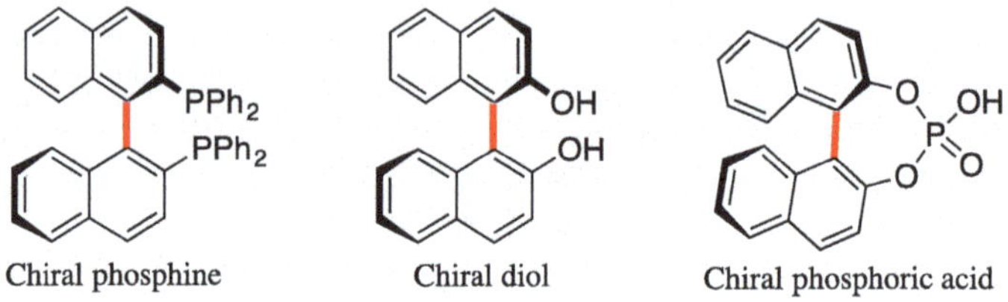

Fig. 1.5 Examples of commonly used atropisomers

The considerable steric hindrance provided by substituents A, B, C, D rises the rotational barrier of the highlighted single bond (which is called a stereogenic axis in these conditions) so the molecule can exist as a pair of stable conformational enantiomers (i.e. atropisomers). Like stereocenters, also stereogenic axes (or chiral axes) have to be somehow labelled and in this elaborate we use the ***P***, ***M*** descriptors (Fig. 1.4).

After assigning the priority to the substituents and to the planes (the plane closer to the observer has the priority),[4] looking at the molecule along the axis, this method considers the rotation of the dihedral angle of the two planes connected by the axis and assigns the ***P*** (plus) configuration for a clockwise rotation and the ***M*** (minus) configuration for a counterclockwise one. Obviously, the ***P*** and ***M*** configurations can only be assigned to stable atropisomers but conformational stability, which is a chemical equilibrium, can be very subjective so it is very important to rule it. Out of the many definitions of this concept, one of the most used is probably that given by Oki [11] stating that atropisomers are defined as axially chiral molecules that can be physically isolated and have a half-life time of interconversion of at least 1000 s (~17 min) at 300 K (27 °C). This is an arbitrary definition but is more efficient compared to others because it takes the temperature into account as it should be for a chemical equilibrium.

Although they are less abundant compared with centrally chiral compounds, atropisomers can still be found in many naturally [12] and synthetically [13] occurring molecules and nowadays play a central role as ligands or catalysts in asymmetric synthesis (Fig. 1.5).

---

[4]The priority of substituents is assigned according to the CIP rules. *Angew. Chem. Int. Ed.*, 1966, 385.

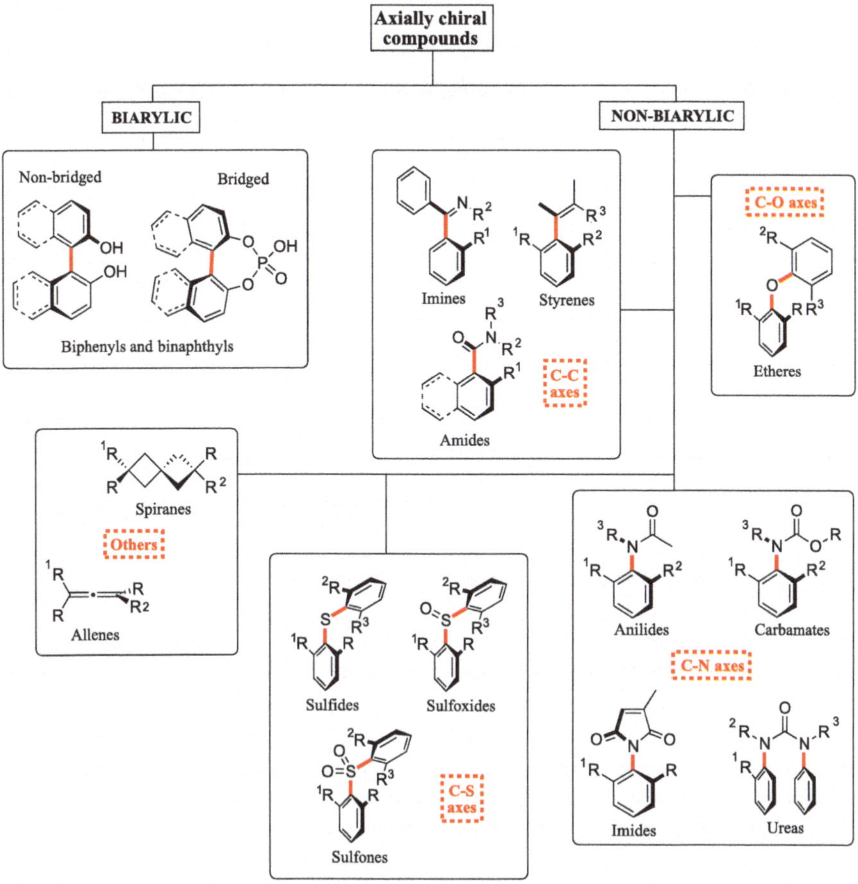

Fig. 1.6 Classes of axially chiral compounds

Figure 1.6 shows some chiral biarylic compounds. These were the first kind of atropisomers ever reported [14] and are by far the most common and studied ones [15]. However, there are many classes of recognized axially chiral compounds [16] (Fig. 1.7) such as amides [17], anilides and ureas [18], imides [19], diaryl ethers, thioethers and sulfones,[5] aryl imines and styrenes [22], *N*-aryl carbazoles and pyrroles [12] and allenes [23].

The ***P-M*** method explained earlier is applicable to all the atropisomers showed in Fig. 1.6, however, molecules like the one in Fig. 1.7 having a C(sp^2)-C(sp^3) stereogenic axis (which are discussed in this elaborate) need additional rules to fully label their conformations.

Considering the substituents on the sp^3 carbon with respect to the sp^2 one, they can exist in many stable conformations. For simplicity let us only consider those showed

[5](a) Betson et al. [20]; (b) Clayden et al. [21].

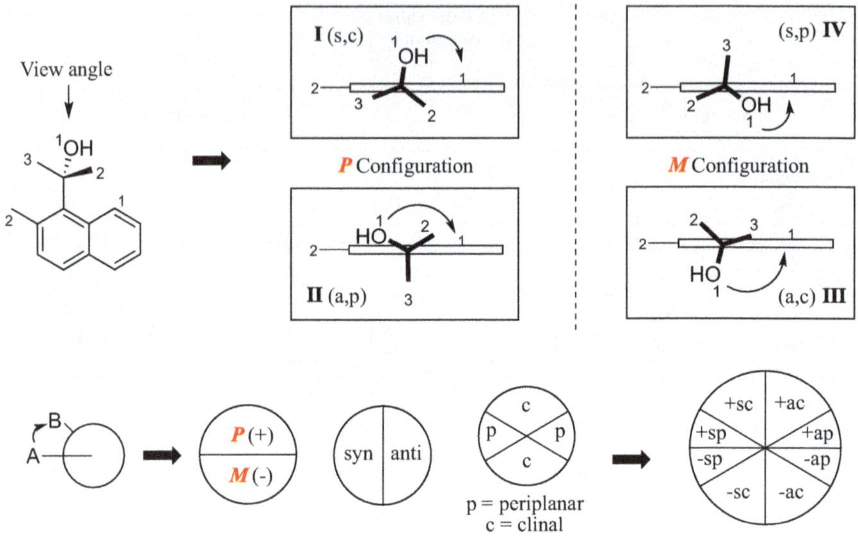

Fig. 1.7 Conformational nomenclature for approximate dihedral angles

above: cases **I-II** and **III-IV** have the same torsion angle (i.e. dihedral angle), positive and negative respectively, but due to restricted rotation are not the same stereoisomer, hence for these molecules the descriptors "syn-anti" and "periplanar-clinal" must be employed. This way the circle is divided in eight portions each labeled with three descriptors (the sign of the dihedral angle, the syn-anti and the periplanar-clinal) in order to name every possible conformation [24].

1.1.3 Helical Chirality

It is a property of screw-shaped objects, the most famous example being the macromolecule of DNA which has the shape of a right-handed double helix (Fig. 1.8).

Simpler molecules also show this kind of chirality and the smallest structure that shows stable helical chirality is called hexahelicene (Fig. 1.9).

The descriptors for helicity are ***P-M*** also in this case, with ***P*** labelling a clockwise rotation and ***M*** a counterclockwise one. There are very few examples of real enantioselective synthesis of helicenes,[6] but an effective one was recently reported by Alcazaro (Reaction 1.3) [28].

[6]For exhaustive reviews on helically chiral compounds see: (a) Shen and Chen [25]; (b) Gingras [26], 968; (c) Gingras et al. [27]; (d) Gingras [26], 1051.

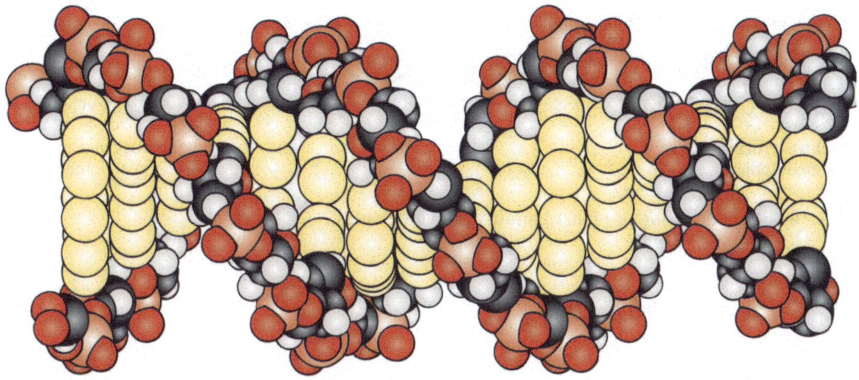

Fig. 1.8 The structure of the DNA

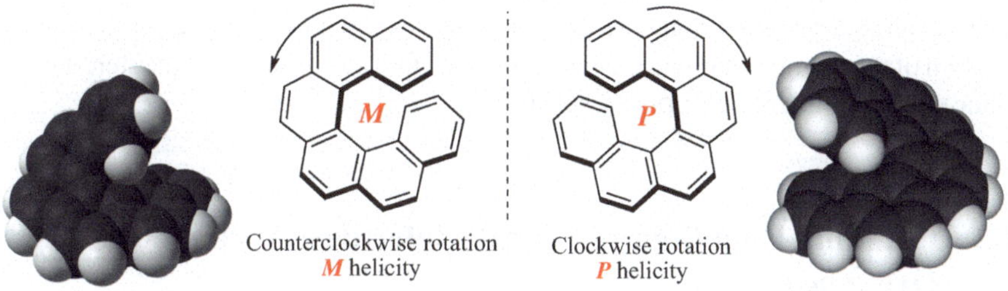

Fig. 1.9 Enantiomers of hexahelicene

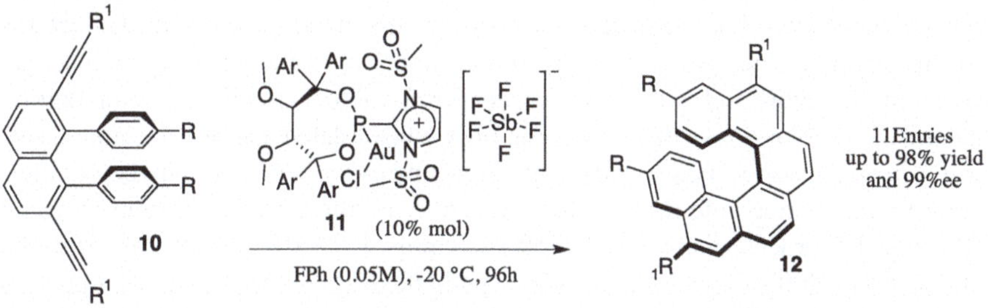

Reaction 1.3 Enantioselective synthesis of hexahelicene

They first built an achiral reagent and then employed the cationic gold phosphinite **11** to activate the alkyne functionalities for the intramolecular hydroarylation reaction. This way they obtained substituted hexahelicene in good yield and selectivity.

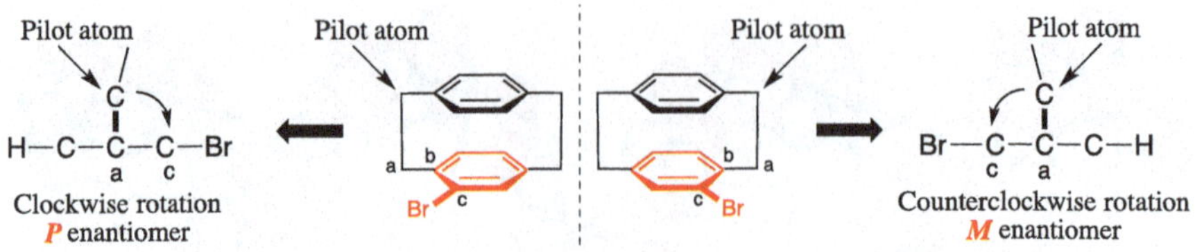

Fig. 1.10 Stereolabelling of planar chirality of cyclophanes

1.1.4 Planar Chirality

Those molecules possessing a stereogenic plane show planar chirality. A plane is termed stereogenic when it cannot lie in a symmetry plane because of a restricted rotation [2]. This kind of chirality is typical of cyclophanes.

The stereodescriptors ***P*** and ***M*** are assigned as shown in Fig. 1.10 basing on the rotation of the plane containing the pilot atom (the first atom outside of the stereogenic plane) towards the highest priority substituent.

1.2 Pursuing Enantiopure Chiral Molecules: Asymmetric Organocatalysis

The concept of molecular chirality is very old and can be traced back to 1875 [29]. Yet it was not fully acknowledged by the pharmaceutical industry until the late 1990's [30]. In this time scientists grew more and more aware of the intrinsic microscopic asymmetry of nature and started considering the possible consequences of marketing racemic drugs. Nowadays this concept is fully consolidated and new racemic drugs must be tested separately as single enantiomers and as a racemic mixture before marketing them, whereas old drugs sold as racemic mixtures are being reinvestigated following that process known as "chiral switching" [31]. Because of this, from the 1980's to the 2000's the number of chiral synthetic drugs sold as single enantiomers rose from 12 to 75% [32]. This incredible progress could not have been possible without the advent of asymmetric synthesis and more importantly asymmetric catalysis. That is the employment of an enantiopure catalyst to generate a large amount of enantioenriched product from a chiral or prochiral reagent. Out of the many ways (i.e. chiral pool, resolution and chromatography) to obtain enantiopure molecules, asymmetric catalysis is certainly the most atom- step- and redox economic one [33], furthermore it is also very challenging and stimulating given the uncountable possibilities it offers and is definitely the right choice for greener and more sustainable processes. This branch of chemistry is commonly divided into three areas that are enzymatic, metal and organic catalysis each with its advantages and disadvantages that make them complementary to each others [34]. The key point of this Ph.D. work has been asymmetric organocatalysis. Its "explosion" has been triggered by two

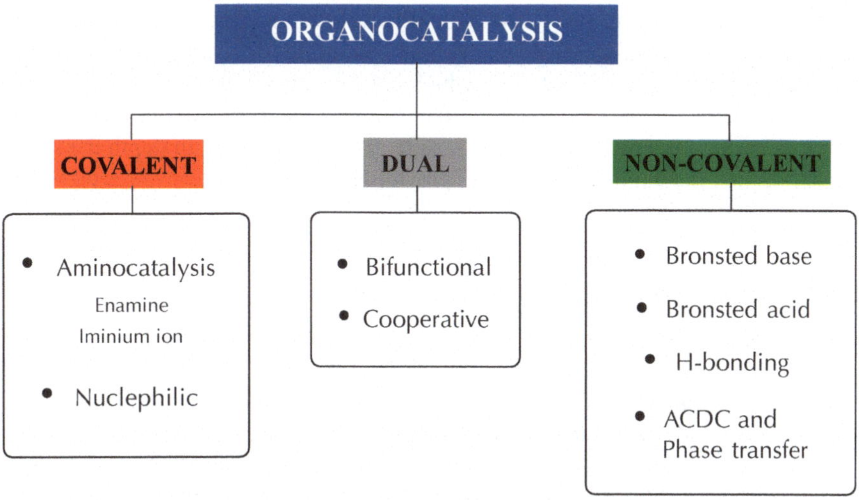

Fig. 1.11 Classification of the main organocatalytic activation modes

parallel publications at the beginning of 2000 [35, 36] by MacMillan and List where this discipline was defined as an independent area of chemistry but more importantly where its broad applicability was demonstrated. Basically, those two reports showed that the catalytic activations employed were not restricted to a single reaction but were absolutely general and could be applied to a plethora of new transformations [37]. Because of this, organocatalysis has experienced an incredible growth and is now a mature and organized branch of chemistry which is usually divided into sub-areas basing on the catalyst and the activation mode it provides.[7] There are as many activation modes as many families of catalysts (Fig. 1.11) and generally the parameter used for their classification is the interaction they establish with the reagent to activate it so that two main categories can be outlined: covalent and non-covalent catalysis.

As the names suggest, the catalyst accelerates the reaction rate by making a covalent or a non-covalent interaction with the substrates. Naturally there are not overall better and worse activation modes and it will be the chemist's job to choose the proper one (which means choosing the proper family of catalysts) for its reaction. Usually though, they have their own features: for example, covalent catalysis shows an overall higher level of stereocontrol because the catalyst is directly bonded with the substrate exerting its chiral influence at the best. Also, given the directional nature of a covalent bond, the stereochemical outcome of a reaction can be predicted and accounted to some extent. On the contrary non-covalent catalysis is based on weaker interactions (e.g. hydrogen bonds, Van der Waals, ion pairing) which are intrinsically undirectional so that it is more difficult to have a good selectivity and to account for it. On the other hand, a weak interaction is more easily established

[7] (a) MacMillan [38]; (b) List [39].

making this catalysis generally faster than the covalent one which means that a lower reaction time or a lower catalyst loading are required. Finally, in Fig. 1.11 there is a third category called "dual" catalysis which indicates that more than one activation mode is occurring. In the next chapters, many of these catalysis like amino, Brønsted base and bifunctional will be presented, so let us examine more in depth the single activations and the families of catalysts providing them.

1.2.1 Aminocatalysis: Enamine and Iminium Ion

It is only natural to explain aminocatalysis with the seminal works by List and MacMillan that launched it [35, 36], but before we get to them, it seems fair to give a bit of a historical background. Using a small chiral amine as a catalyst is nowadays an established and efficient routine in modern asymmetric synthesis,[8] but its origins are dated back to 1896 when the German chemist Emil Knoevenagel reported the first achiral aminocatalytic aldol condensation reaction that takes after his name (Reaction 1.4) [40].

Knoevenagel had already understood that only a catalytic amount of achiral piperidine **15** was needed leading the way for the modern asymmetric application of our time to the point that List himself admitted [41] there is a direct connection between Knoevenagel's and his own (and probably also MacMillan's) seminal work. So let us start with enamine catalysis and see the reaction reported by List (Reaction 1.5).

It is an intermolecular aldol reaction between acetone and various aldehydes. The central point of this strategy is to form the enamine of the acetone to enhance its nucleophilicity and promote the attack on the aldehyde. In his paper, List also proposed a reaction mechanism to account for the selectivity observed (Fig. 1.12).

At the very beginning of the cycle the proline attack on acetone is favored by its higher concentration with respect to the aldehyde and in step (b), after dehydration, the equilibrium between iminium ion and enamine is established (red and blue squares). Next is the step that accounts for the selectivity in which the enamine and

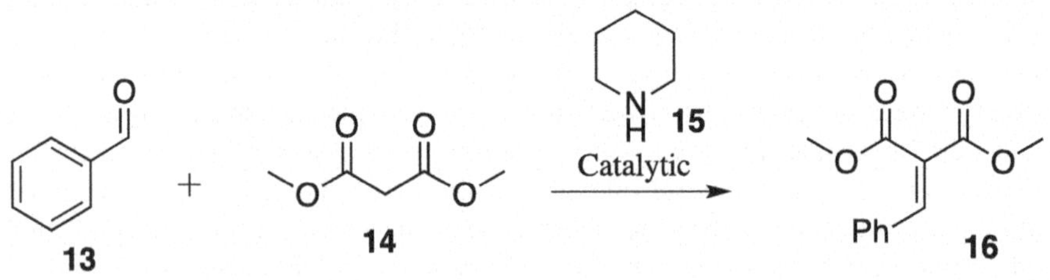

Reaction 1.4 The first Knoevenagel reaction

[8]For exhaustive reading on asymmetric aminocatalysis on both enamine and iminium ion reactions see: Science of synthesis "Asymmetric organocatalysis 1—Lewis base and acid catalysis", *Thieme*, 2012.

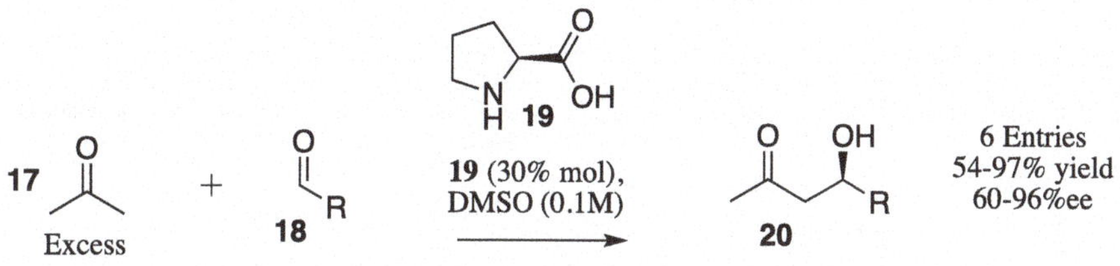

Reaction 1.5 The proline-catalyzed enantioselective intermolecular aldol reaction

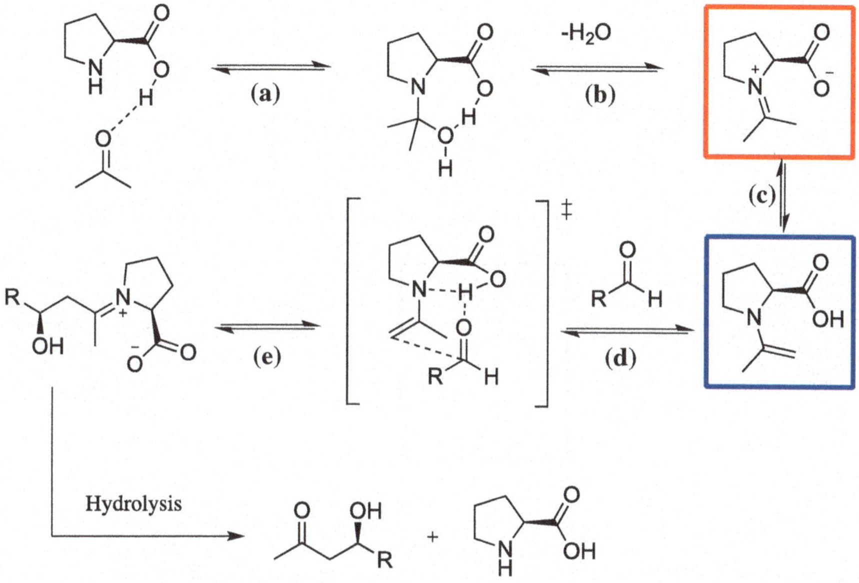

Fig. 1.12 Proposed enamine mechanism

aldehyde form a tricyclic intermediate, favored by a network of hydrogen bonds, that brings the two substrates close enough so that the selective attack can take place in step (e) leaving only hydrolysis left to restore the catalyst and afford the enantioenriched product. An important aspect of this catalytic cycle is the equilibrium between the enamine and the corresponding iminium ion. As List states in a later publication [39] these two species are closely related and yet are electronically opposite so they can be used as opposite synthons in a chemical transformation with the first, enhancing the nucleophilic properties of a carbonyl compound and the latter, enhancing its electrophilic ones. The job of the chemist is to find the right conditions and the right reaction partners in order to drive the equilibrium towards only one of the two species which is exactly what happens in MacMillan work. He reported the organocatalytic intermolecular DA reaction between α,β-unsaturated aldehydes activated via iminium ion and various dienophiles (Reaction 1.6).

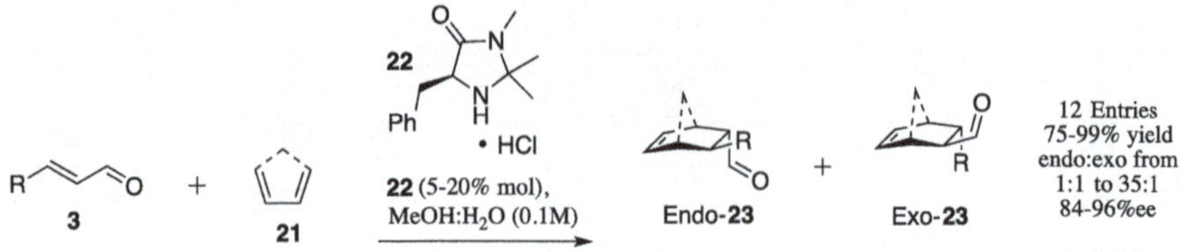

Reaction 1.6 The asymmetric DA reaction of iminium ion activated dienophiles

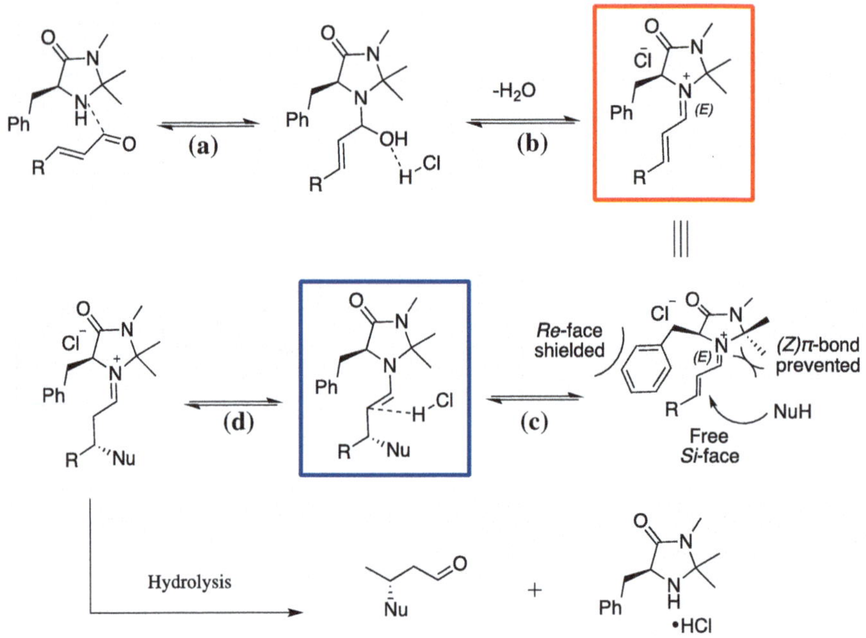

Fig. 1.13 Iminium ion mechanism

In this case MacMillan did not draw a detailed mechanism but declared which were the most important and stereoinducting steps and backed his hypothesis up with calculations so it is possible to draw a mechanism (Fig. 1.13).[9]

Once again, the cycle begins with the formation of the iminium ion (in the red square). The dimethyl substituents on the right prevent the formation of the *Z* double bond so the electrophilic carbon is forced below the plane of the phenyl ring which efficiently shields the *Re*-face. The nucleophilic attack in step (c) produces an enamine (in the blue square) that is protonated and hydrolyzed to restore the catalyst and yield the enantioenriched product. As I mentioned previously, enamine and iminium ion are closely related and it is easy to see how they are present in each other

[9]For a better understanding of the concept, instead of drawing two mechanisms for the *endo* and *exo* products, a single one is reported with a generic nucleophile "NuH". Also, the enantiotopic faces of the iminium ion may change depending on the nature of the R substituent.

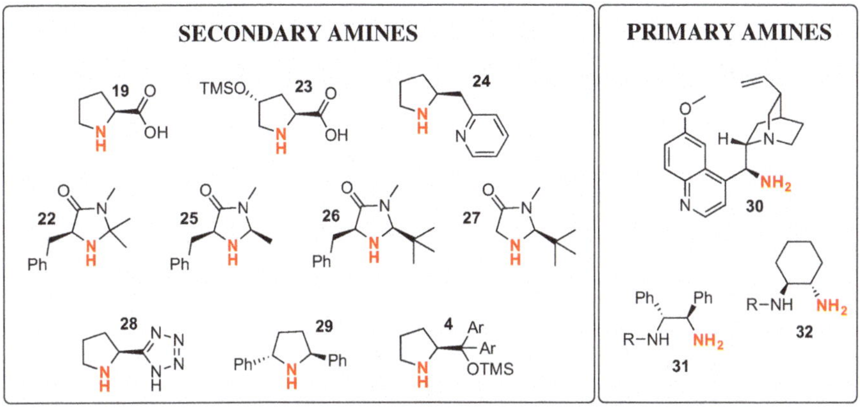

Fig. 1.14 Some representative secondary and primary chiral amines

catalytic cycles which ultimately leads to the development of cascade reactions like Reactions 1.1 and 1.2 where one catalyst accelerates more than one transformation one-pot.

Finally, it is very important to notice that both reactions were performed with cheap, stable organic catalysts (Fig. 1.14) without exclusion of air and moisture from the reaction medium.

This is their greatest advantage and was revolutionary in a time where metals dominated the stage. In contrast to metal catalysis, the use of mild conditions together with its broad applicability contributed to the incredible development of aminocatalysis (and organocatalysis in general) as proven by the huge number of papers that followed these pioneering works.

1.2.2 Nucleophilic Catalysis

This branch of catalysis is based on the ability of a chiral nucleophile to activate an electron-poor species by making a reversible covalent bond with it. Some typical examples of this reactivity are the NHC-catalyzed benzoin condensation (Fig. 1.15a)[10] and the phosphine- or tertiary amine-catalyzed MBH reaction (Fig. 1.15b).[11]

Intermediate **III** of the NHC catalytic cycle is basically an enamine whereas in the MBH reaction, zwitterionic intermediate **II** is basically an enolate. One can easily see the usefulness of this catalysis enabling an electron-poor compound to act as a nucleophile especially in the first case where "umpolung" reactivity [49] is achieved

[10]For reviews on NHC-catalysis see: (a) Biju et al. [42]; (b) Grossman and Enders [43]; (c) Chen and Ye [44]; (d) Hopkinson et al. [45]; (e) Wang and Scheidt [46].

[11]For reviews on MBH and MBH-like reactions see: (a) Wei and Shi [47]; (b) Rios [48].

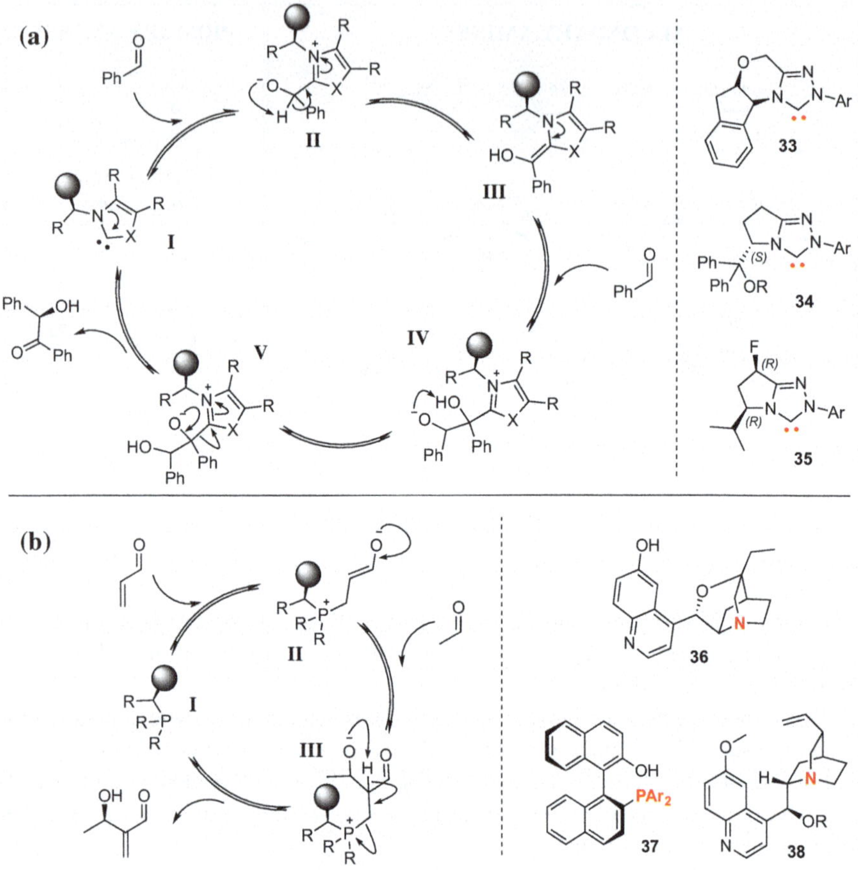

Fig. 1.5 **a** NHC general catalysis and catalysts; **b** MBH reaction and catalysts

and the aldehyde carbon is turned nucleophilic without resorting to multistep procedures involving stoichiometric (and very unpleasant) dithiols. A recent example of an NHC-catalyzed reaction was reported by Glorius and collaborators (Reaction 1.7) [50].

They skillfully employed this reactivity to access optically active complex spirocyclic structures. Chiral carbenes **40** and **43** efficiently promote the *umpolung* of the aldehyde. After protonation of the heterocyclic double bond, the nucleophilic carbon selectively attacks the *Si* face of the furan and thiophene moieties to afford the product in very good yield and enantioselectivity.

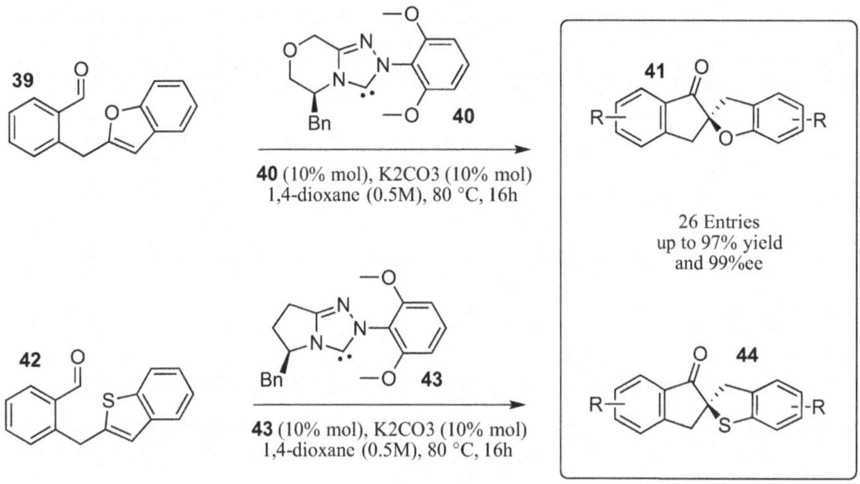

Reaction 1.7 NHC-catalized umpolung of aldehyde for the construction of chiral spirocycles

1.2.3 Brønsted Acid and Base Catalysis[12]

Throughout all areas of chemistry, the majority of the synthetic processes relay on acid-base catalysis [52]. Be it homogeneous or heterogeneous, large- or laboratory scale, there are legions of reactions favored by an acidic or a basic catalyst. The simple aldol condensation between acetone and benzaldehyde or the Fisher esterification are catalyzed by KOH and AcOH respectively. Non-covalent catalysis can be applied to asymmetric organocatalysis too with the same concept we have seen until now: a chiral enantiopure catalyst is used so that it does not only promote the reaction but also induces chirality in the products. The most used Brønsted acids and bases are by far BINOL-derived phosphoric acids, its spirocyclic analogues and *N*-phosphoramides (Fig. 1.16a) and cinchona alkaloid-derived tertiary amines (Fig. 1.16b).

The *modus operandi* of a Brønsted acid/base is to activate a substrate by donating/accepting protons generating partially or completely charged intermediates in order to form an ion pair with the chiral catalyst and to undergo an enantioselective transformation. Although in many reactions the catalysis proceeds this way, there are also some reported examples where the protonation/deprotonation is the stereoinducting step of the reaction [53].

One recent example of Brønsted acid catalysis reported by List [54]. shows a new and unusual phosphoric acid dimer promoting an oxa-Pictet-Spengler reaction (Reaction 1.8).

[12](a) For an exhaustive reading on Brønsted acid/base catalysis see: Rueping et al. [51]; (b) Science of synthesis "Asymmetric organocatalysis 2 – Brønsted base and acid catalysis, and additional topics", Thieme, 2012.

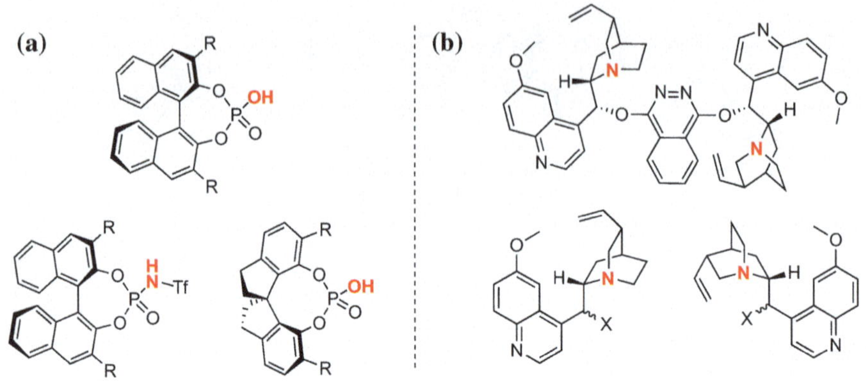

Fig. 1.16 **a** BINOL-derived Brønsted acid; **b** cinchona alkaloid-derived tertiary amines

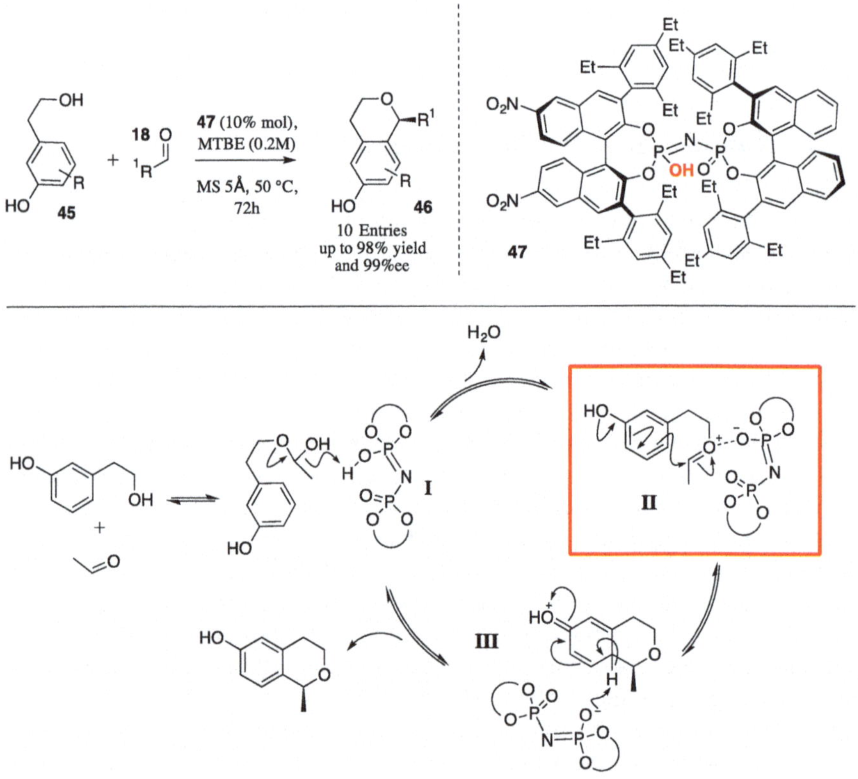

Reaction 1.8 The asymmetric oxa-Pictet-Spengler reaction catalyzed by a PA dimer

The alkyl alcohol condenses on the aldehyde to form a hemiacetal which is dehydrated by the acid (step **I**) to form an oxonium ion that is the key intermediate of the

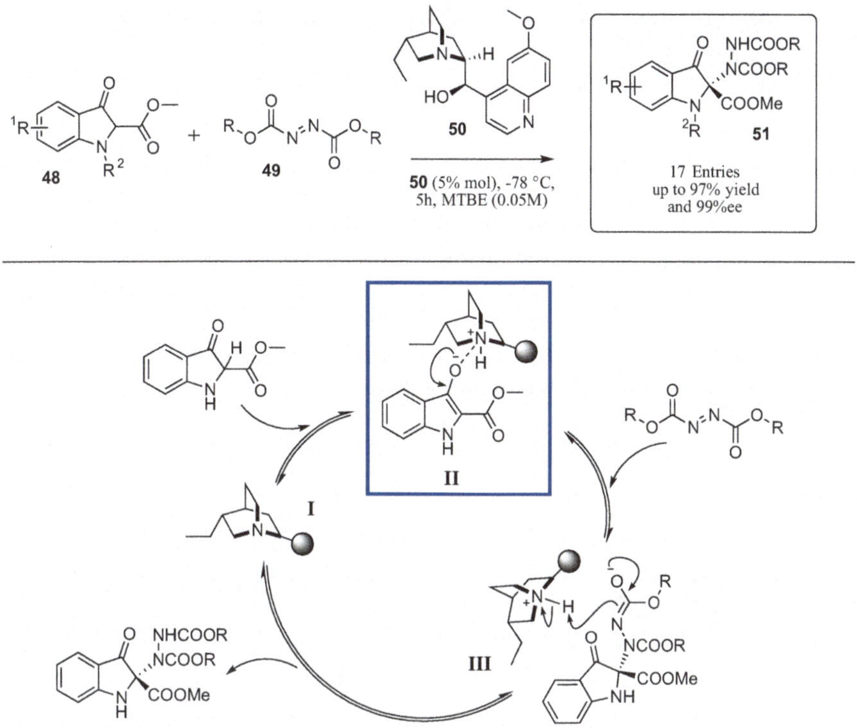

Reaction 1.9 The asymmetric amination of indolin-3-ones

reaction because it leads to the formation of the new C-C bond. This is formed in a stereoselective fashion because of the ion pairing with the phosphate that forces the reagent to adapt itself to the chiral environment of the catalyst leading to the selective *Si* face attack on the oxonium double bond. In the final step, rearomatization restores the acid and affords the product in very good yield and enantioselectivity.

Brønsted base catalysis follows the same principles but with opposite polarity because the catalyst now takes a proton and becomes positively charged. Reddy and coworkers have recently reported an example of a Brønsted base-catalyzed amination reaction (Reaction 1.9) [55].

By deprotonation, the catalyst promotes the formation of a nucleophilic enolate which forms an ion pair with the resulting ammonium cation. Once again, because of this interaction, the catalyst can exert its chiral influence on the reagents forcing them to approach each other in a selective manner. The final reprotonation step restores the catalyst and affords products with a congested tetrasubstituted carbon in high yield and selectivity.

1.2.4 *Asymmetric Counterion-Directed and Phase-Transfer Catalysis*

We have seen in the previous paragraph the importance of the ion pairing between reagent and catalyst. These species become (partially or completely) charged after protonation/deprotonation by the chiral acid/base which is restored as a neutral compound at the end of its cycle acting as both the catalyst and the stereoinducting agent of the reaction. *ACDC* is a very similar type of catalysis but in this case the catalyst is an achiral compound that generates a charged reactive intermediate which is "captured" by a chiral counterion that acts as the stereoinducting agent.[13] This catalysis was first reported and defined by List in 2006 [58]. In his paper he shows the asymmetric hydrogenation of α,β-unsaturated aldehydes with a modified Hantzsch ester (Reaction 1.10).

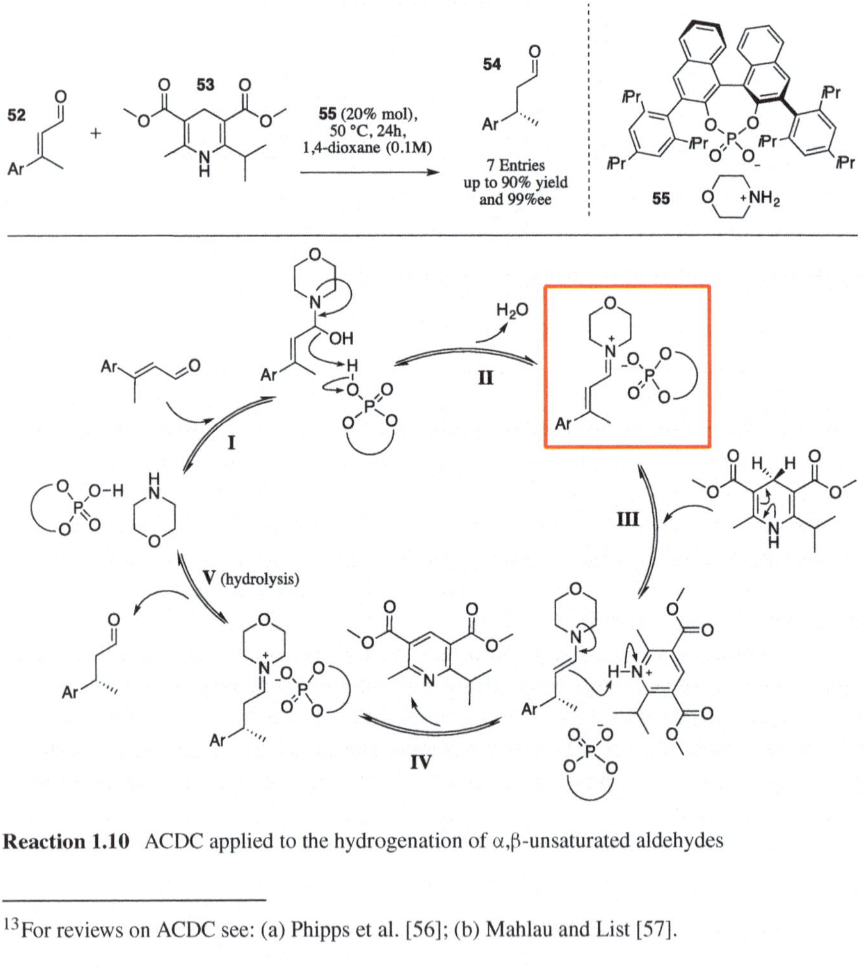

Reaction 1.10 ACDC applied to the hydrogenation of α,β-unsaturated aldehydes

[13]For reviews on ACDC see: (a) Phipps et al. [56]; (b) Mahlau and List [57].

The cycle starts with the condensation of morpholine on the aldehyde to generate the *achiral* iminium ion whose counter ion is the atropisomeric phosphate (in the red square). It is evident from this example how the generation of the reactive intermediate is promoted by an achiral compound whereas the enantioselectivity can only be provided by the chiral anion that coordinates the iminium ion so that the hydride addition happens exclusively on the *Re* face of the activated double bond. At this point the stereochemistry of the system has been defined and further protonation and hydrolysis restore the catalytic salt and afford the product in high yield and enantioselectivity. This kind of catalysis can naturally be performed also with a positively charged counterion which is usually (but not necessarily) a chiral quaternary ammonium ion and the concept of *ACDC* has also been extended to biphasic reactions and is referred to as phase-transfer catalysis.[14] The mechanism is exemplified in Fig. 1.17a[15] and, except for the two layers, is identical to the previous case: an external compound, typically an inorganic acid/base soluble in the aqueous layer, generates a charged intermediate at the interface which undergoes an ion exchange with the chiral PT catalyst (the quaternary ammonium). Then the catalyst carries the charged intermediate in the organic layer where the displacement reaction takes place (in a stereoselective manner thanks to the ion-pairing) and the catalyst is restored.

PTC has found many applications also on industrial scale [64] and above (Fig. 1.17b) are reported some of the most used catalysts both cationic and anionic.

1.2.5 Hydrogen-Bonding Catalysis[16]

For a long time, H-bonding interactions have been labeled as too weak to be able to efficiently promote a reaction, be it asymmetric or not, until pioneering studies by Jacobsen and Corey[17] on hydrocyanation of imines demonstrated the opposite (Reaction 1.11a, b).

Supporting the experimental results with calculations, they showed how a chiral H-bond donor can not only activate an EWG but also promote its reaction with high enantioselectivity. Having said that, it is relatively rare to find reactions that are catalyzed exclusively by H-bonding interactions, whereas it is very common to find this activation paired with another one in dual catalytic processes that are discussed in the next paragraph.

[14]For exhaustive reviews on PTC see: (a) Ooi and Maruoka [59]; (b) Shirakawa and Maruoka [60]; (c) Albanese et al. [61].

[15]Asymmetric PTC is believed to mainly follow the "interfacial mechanism" which is the only one reported for simplicity. See: Kitamura et al. [62]. For reading on the "extraction mechanism" see: Starks [63].

[16]For reviews see: (a) Doyle and Jacobsen [65]; (b) Knowles and Jacobsen [66].

[17](a) Sigman and Jacobsen [67]; (b) Corey and Grogan [68].

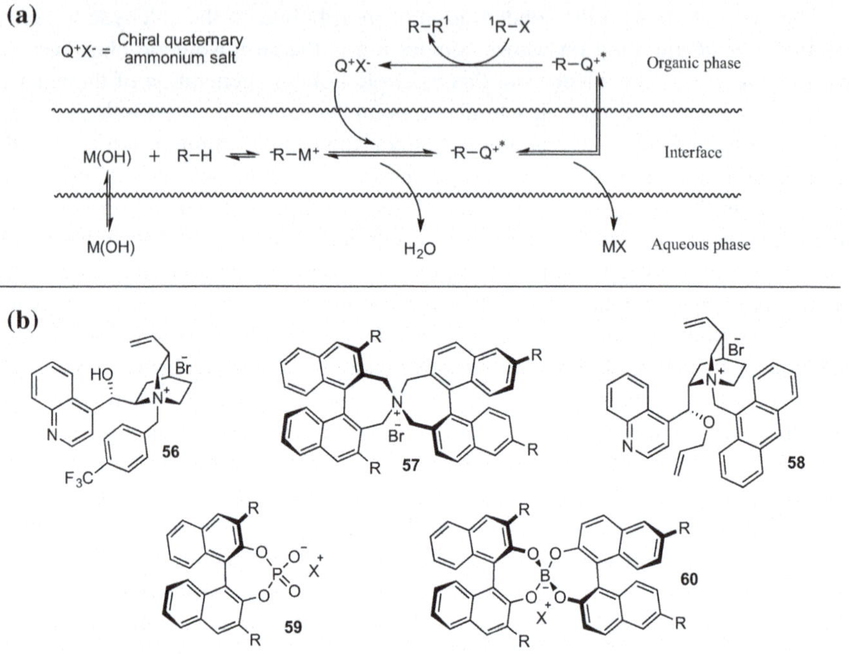

Fig. 1.17 **a** The general PTC mechanism; **b** cinchona alkaloid- and BINOL-derived counterion catalysts

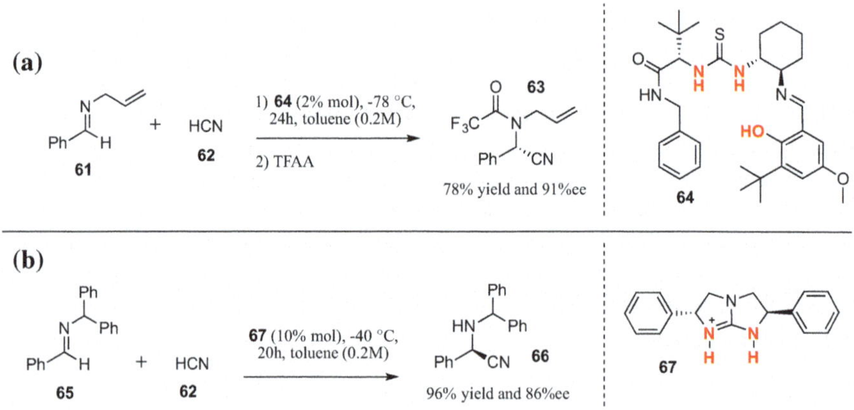

Reaction 1.11 Hydrocyanation of imines by **a** Jacobsen; **b** Corey

1.2.6 Dual Catalysis: Bifunctional and Cooperative

It is the simultaneous employment of two activation modes in a single reaction.[18] It can be bifunctional or cooperative and the difference between them is that in the first case the double activation is provided by the same catalyst, whereas in the latter case there are two or more catalysts (Fig. 1.18).

Although it is not an independent type of catalysis itself, it is worth mentioning it because there are many transformations that can only be promoted by dual catalysis both in terms of reactivity and selectivity. Many applications of this concept, not necessarily only organocatalytic ones, have been reported by many groups[19] showing, for example, how organic and metal catalysis can be merged together to access some innovative transformations like the direct β-functionalization of saturated carbonyl compounds by MacMillan (Reaction 1.12a) [78].

The authors also proposed a mechanism (Reaction 1.12b) for this reaction. The organocatalytic cycle starts with the classical enamine formation and intercepts the photoredox cycle in a SET process where the Ir^{IV} is reduced to Ir^{III}. This species can now be excited by light again and reduces the cyanobenzene substrate in the other SET step affording the aryl anion, which is a relatively stable radical, and the

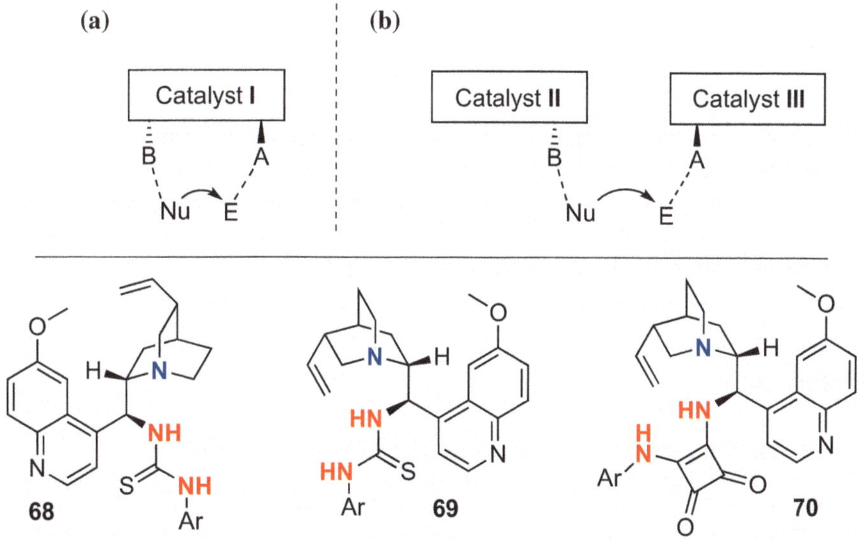

Fig. 1.18 **a** Bifunctional and **b** cooperative catalysis and some of the most common bifunctional catalysts

[18]For reviews see: (a) Shao and Zhang [69]; (b) Zhong and Shi [70]; (c) Allen and MacMillan [71]; (d) Du [72]; (e) Afewerki and Còrdova [73].

[19]For some remarkable examples of dual catalysis see: (a) Krautwald et al. [74]; (b) Rono et al. [75]; (c) Noesborg et al. [76]; (d) Meazza et al. [77].

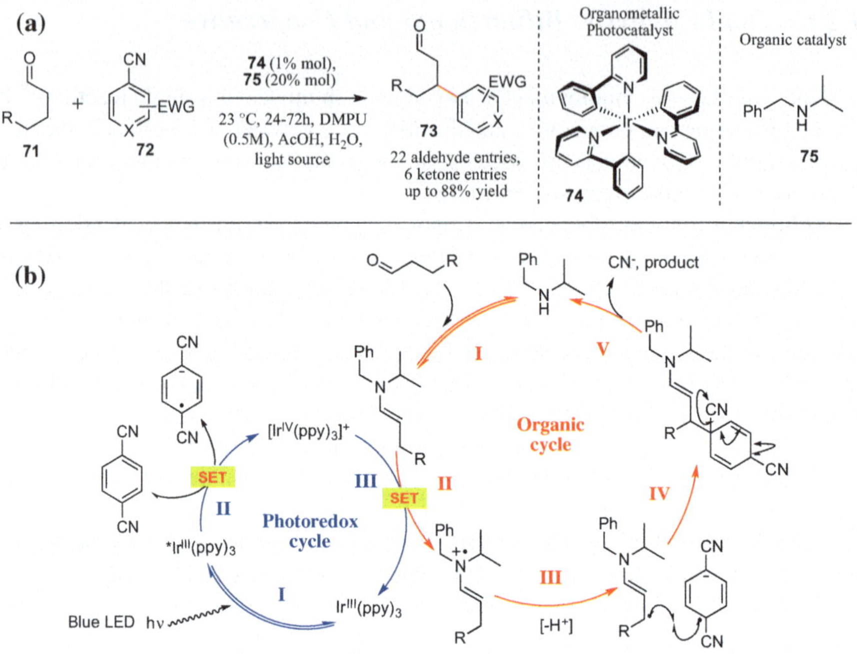

Reaction 1.12 **a** The direct β-functionalization of carbonyl compounds; **b** proposed mechanism

Ir^{IV} complex. Being extremely oxidant, it takes an electron from the electron-rich enamine to generate an enaminyl radical cation that after deprotonation couples with the aryl cyanide and affords the products. This remarkable transformation can only be achieved with the simultaneous contribution of organic and photoredox catalysis showing how powerful a dual catalytic process can be.

References

1. Eliel EL, Wilen SH (1994) Stereochemistry of organic compounds. Wiley
2. IUPAC (1997) Compendium of chemical terminology, 2nd edn. (the "Gold Book"). Oxford
3. Wolf C (2008) Dynamic stereochemistry of chiral compounds: principles and applications RSC, p 71
4. Enders D, Hüttl MRM, Grondal C, Raabe G (2006) Nature 861
5. Enders D, Hüttl MRM, Runsink J, Raabe G, Wendt B (2007) Angew Chem Int Ed 467
6. Enders D, Hüttl RM, Raabe G, Bats JW (2008) Adv Synth Catal 267
7. Cassani C, Tian X, Escudero-Adàn EC, Melchiorre P (2011) Chem Commun 233
8. Zhou B, Yang Y, Shi J, Luo Z, Li Y (2013) J Org Chem 2897
9. Chauhan P, Mahajan S, Raabe G, Enders D (2015) Chem Commun 2270
10. Reyes E, Jiang H, Milelli A, Helsner P, Hazell RG, Jorgensen KA (2007) Angew Chem Int Ed 9202
11. Ōki M (1984) Top Stereochem 1–81

12. Smyth JE, Butler NM, Keller PA (2015) Nat Prod Rep 1562
13. Bringmann G, Mortimer AJP, Keller PA, Gresser MJ, Garner J, Breuning M (2005) Angew Chem Int Ed 5384
14. Christie GH, Kenner J (1922) J Chem Soc Trans 614
15. Bringmann G, Gulder T, Gulder TAM, Breuning M (2011) Chem Rev 563
16. Kumarasamy E, Raghunathan R, Sibi MP, Sivaguru J (2015) Chem Rev 11239
17. Clayden JP, Lai LW (1999) Angew Chem Int Ed 2556
18. Adler T, Bonjoch J, Clayden J, Font-Bardfa M, Pickworth M, Solans X, Sole D, Vallverdu L (2005) Org Biomol Chem 3173
19. Curran DP, Qi H, Geib SJ, DeMello NC (1994) J Am Chem Soc 3131
20. Betson MS, Clayden J, Worrall CP, Peace S (2006) Angew Chem Int Ed 5803
21. Clayden J, Senior J, Helliwell M (2009) Angew Chem Int Ed 6270
22. Pinkus AG, Riggs JI, Broughton SM (1968) J Am Chem Soc 5043
23. Taylor DR (1967) Chem Rev 317
24. Klyne W, Prelog V (1960) Experientia 521
25. Shen Y, Chen C (2011) Chem Rev 1463
26. Gingras M (2013) Chem Soc Rev
27. Gingras M, Félix G, Peresutti R (2013) Chem Soc Rev 1007
28. González-Fernández E, Nicholls LDM, Schaaf LD, Farès C, Lehmann CW, Alcazaro M (2017) J Am Chem Soc. ASAP article. https://doi.org/10.1021/jacs.6b12443
29. van't Hoff JH (1875) Bull Soc Chim France 295
30. Hutt AJ, O'grady J (1996) J Antimicrob Chemoter 7
31. Agranat I, Caner H, Caldwell J (2002) Nat Rev Drug Discov 753
32. Aliens EJ, Wuis EW, Veringa EJ (1988) Biochem Pharmacol 9
33. Burns NZ, Baran PS, Hoffmann RW (2009) Angew Chem Int Ed 2854
34. List B (2004) Adv Synth Catal 1021
35. Ahrendt KA, Borths CJ, MacMillan DWC (2000) J Am Chem Soc 4243
36. List B, Lerner RA, Barbas III CF (2000) J Am Chem Soc 2395
37. Gaunt MJ, Johansson CCC, McNally A, Vo NT (2007) Drug Discov Today 8
38. MacMillan DWC (2008) Nature 304
39. List B (2006) Chem Commun 819
40. Knoevenagel E (1896) Ber Dtsch Chem Ges 172
41. List B (2010) Angew Chem Int Ed 1730
42. Biju AT, Kuhl N, Glorius F (2011) Acc Chem Res 1182
43. Grossman A, Enders D (2012) Angew Chem Int Ed 314
44. Chen XY, Ye S (2013) Org Biomol Chem 7991
45. Hopkinson MN, Richter C, Schedler M, Glorius F (2014) Nature 485
46. Wang MH, Scheidt KA (2016) Angew Chem Int Ed 14912
47. Wei Y, Shi M (2013) Chem Rev 6659
48. Rios R (2012) Catal Sci Technol 267
49. Bugaut X, Glorius F (2012) Chem Soc Rev 3511
50. Janssen-Müller D, Fleige M, Schlüns D, Wollenburg M, Daniliuc CG, Neugebauer J, Glorius F (2016) ACS Catal 5735
51. Rueping M, Parmar D, Sugimoto E (2016) Asymmetric Brønsted acid catalysis. Wiley-VCH Verlag GmbH & Co. KGaA, Weinheim, Germany
52. Parmar D, Sugiono E, Raja S, Rueping M (2014) Chem Rev 9047
53. Lee J-W, List B (2012) J Am Chem Soc 18245
54. Das S, Liu L, Zheng Y, Alachraf MW, Thiel W, De CK, List B (2016) J Am Chem Soc 9429
55. Yarlagadda S, Ramesh B, Reddy CR, Srinivas L, Sridhar B, Reddy BVS (2017) Org Lett 170
56. Phipps RJ, Hamilton GL, Toste DF (2012) Nature Chem 603
57. Mahlau M, List B (2013) Angew Chem Int Ed 518
58. Mayer S, List (2006) Angew Chem Int Ed 4193
59. Ooi T, Maruoka K (2007) Angew Chem Int 4222
60. Shirakawa S, Maruoka K (2013) Angew Chem Int Ed 4312

61. Albanese DCM, Foschi F, Penso M (2016) Org Process Res Dev 129
62. Kitamura M, Shirakawa S, Maruoka K (2005) Angew Chem Int Ed 1549
63. Starks CM (1971) J Am Chem Soc 195
64. Tan J, Yasuda N (2015) Org Process Res Dev 1731
65. Doyle AG, Jacobsen EN (2007) Chem Rev 5713
66. Knowles RR, Jacobsen EN (2010) Proc Natl Acad Sci 20678
67. Sigman MS, Jacobsen EN (1998) J Am Chem Soc 4901
68. Corey EJ, Grogan MJ (1999) Org Lett 157
69. Shao Z, Zhang H (2009) Chem Soc Rev 2745
70. Zhong C, Shi X (2010) Eur J Org Chem 2999
71. Allen AE, MacMillan DWC (2012) Chem Sci 633
72. Du Z, Shao Z (2013) Chem Soc Rev 1337
73. Afewerki S, Còrdova A (20160) Chem Rev 13512
74. Krautwald S, Sarlah D, Schfroth MA, Carreira EM (2013) Science 1065
75. Rono LJ, Yayla HG, Wang DY, Armstrong MF, Knowles RR (2013) J Am Chem Soc 17735
76. Noesborg L, Halskov KS, Tur F, Mønsted SMN, Jørgensen KA (2015) Angew Chem Int Ed 10193
77. Meazza M, Tur F, Hammer N, Jørgensen KA (2017) Angew Chem Int Ed 1634
78. Pirnot MT, Rankic DA, Martin DBC, MacMillan DWC (2013) Science 1593

Chapter 2
The Vinylogous Reactivity of Oxindoles Bearing Non-symmetric 3-Alkylidene Groups

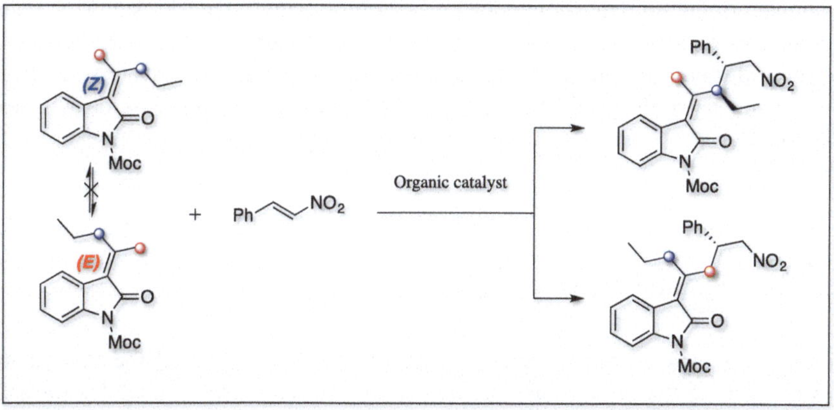

Parts of this chapter were adapted from "N. Di Iorio, P. Righi, S. Ranieri, A. Mazzanti, R. G. Margutta, G. Bencivenni *J. Org. Chem.*, **2015**, 7158" with permission from JOURNAL of ORGANIC CHEMISTRY. Copyrights 2015 American Chemical Society.

Original content can be found at http://pubs.acs.org/doi/abs/10.1021/acs.joc.5b01022.

[1]Fuson [1].

Published as: **N. Di Iorio**, P. Righi, S. Ranieri, A. Mazzanti, R. G. Margutta, G. Bencivenni *J. Org. Chem.*, **2015**, 7158.

N. Di Iorio, *New Organocatalytic Strategies for the Selective Synthesis of Centrally and Axially Chiral Molecules*, Springer Theses,
https://doi.org/10.1007/978-3-319-74914-3_2

Fig. 2.1 Effects of vinylogy

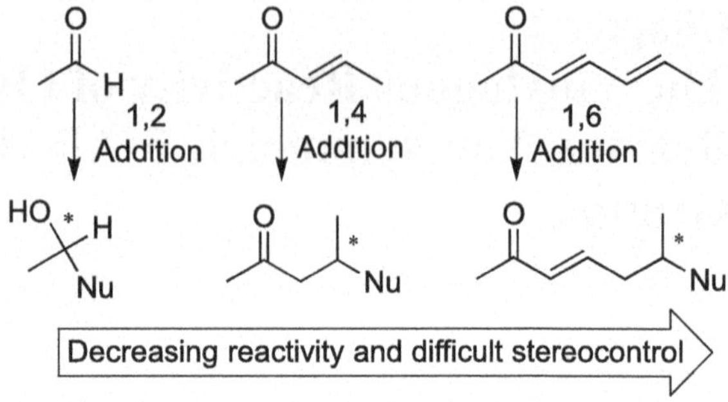

2.1 Vinylogy

Vinylogy, as defined by Fuson,[1] is the possibility to propagate the properties of a functional group through a conjugated unsaturation (vinyl). Under a synthetic point of view, this phenomenon gives access to new transformations that, as a drawback, are more difficult to achieve because the reagent is usually less reactive and the stereochemistry of the product is harder to control as a consequence of the activated functional group being more distant from where the reaction actually takes place (Fig. 2.1).[2]

2.2 Results and Discussion

During the course of our research, Casiraghi and Wang reported the first examples concerning the organocatalyzed reactivity of 3-alkylidene oxindoles towards nitroolefins (Reaction 2.1).[3]

The Michael addition proceeds in the presence of bifunctional catalysts **79** (DHQA-TU) and **83** (QDA-TU), and although the high yields and selectivity achieved, there are still some critical aspects for this reaction that remained unknown and untouched by the authors. First, in every case the presence of an *s-cis* productive dienolate intermediate is proposed without giving any experimental evidence of that and second, the reaction is only performed with substrates possessing only one vinylogous position or two equivalent ones (Fig. 2.2).

That is why we decided to investigate this reaction with oxindoles having two non-equivalent vinylogous positions, γ and γ' (γ has arbitrarily been assigned to the position cis with respect to the carbonyl oxygen). The first issue that we had to face while dealing with these systems is the instability of the double bond, in fact the two isomers interconvert into each other under basic conditions and in 24 h afford the same equilibrium mixture of **Z:E** in a 66:34 ratio.

[2]For further reading on vinylogy see: (a) Pansare and Paul [2]; (b) Hepburn et al. [3].

[3](a) Curti et al. [4]; (b) Rassu et al. [5]; (c) Chen et al. [6].

It is evident from Fig. 2.3 how problematic this isomerization is because there are now two possible *s-cis* intermediates increasing the stereochemical complexity of the reaction. If we take into account this aspect and consider every possible nucleophilic attack (also from the α position) to the nitroolefin acceptor, there is a total of 24 isomers that can be formed in this reaction (Fig. 2.4).

We took on the challenges of this Michael addition aiming to promote it in a regio-, diastereo- and enantioselective manner and to propose a reasonable reaction mechanism, based on experimental evidence, that accounts for the selectivity observed.

We first tried to find the optimal conditions for this reaction starting from stereopure E-oxindole **84**. What we saw is the formation of a mixture of products **86** and **87**, both functionalized in γ, that is the same mixture we observed when we separately reacted stereopure Z-oxindole **85** (Table 2.1).

This outcome is clearly due to the previously mentioned isomerization of the reagents so we figured that a lower temperature would help inhibit the interconversion between the two isomers. After a short temperature screening, we confirmed our hypothesis and observed better results with lower temperature until at −20 °C we obtained complete regio- diastereo- and enantioselectivity for both reagents **84** and **85**. At this point we wanted to investigate the origin of such high selectivity and to do so we performed a series of experiments with deuterated substrates. Basically,

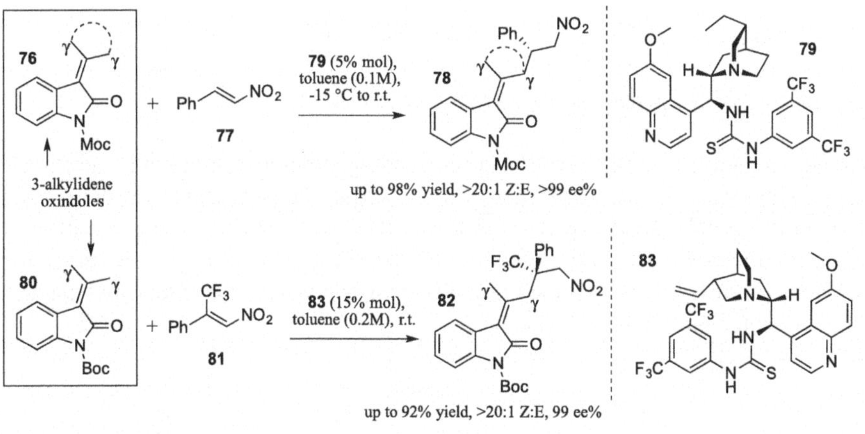

Reaction 2.1 Reactions of 3-alkylidene oxindoles with nitroolefins

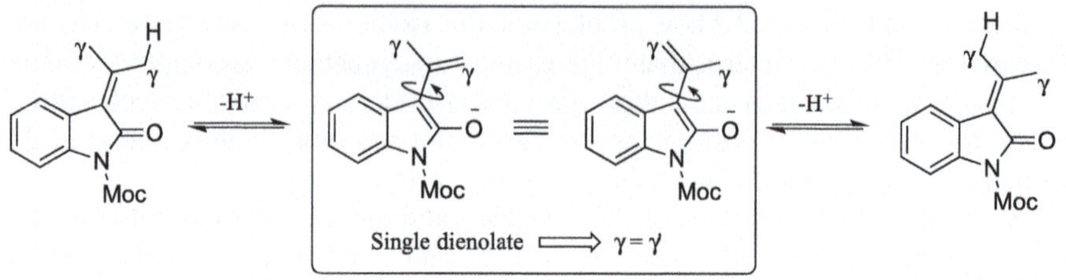

Fig. 2.2 Equivalence of the two vinylogous positions

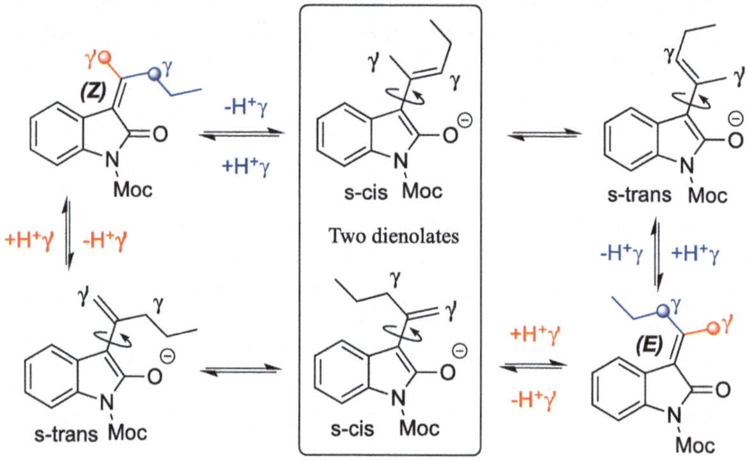

Fig. 2.3 The base-catalyzed isomerization of E/Z oxindoles

for both oxindoles we prepared the corresponding structures completely deuterated at the vinylogous positions (**89** and **93**) and those deuterated only on the methyl group (**88** and **92**). This way, if we observed KIE, we could say in the first place if the deprotonation is the RDS of the reaction, but more importantly if the catalyst is selective towards γ or γ′ or if the high selectivity is just a matter of thermodynamics of the double bond (Scheme 2.1).

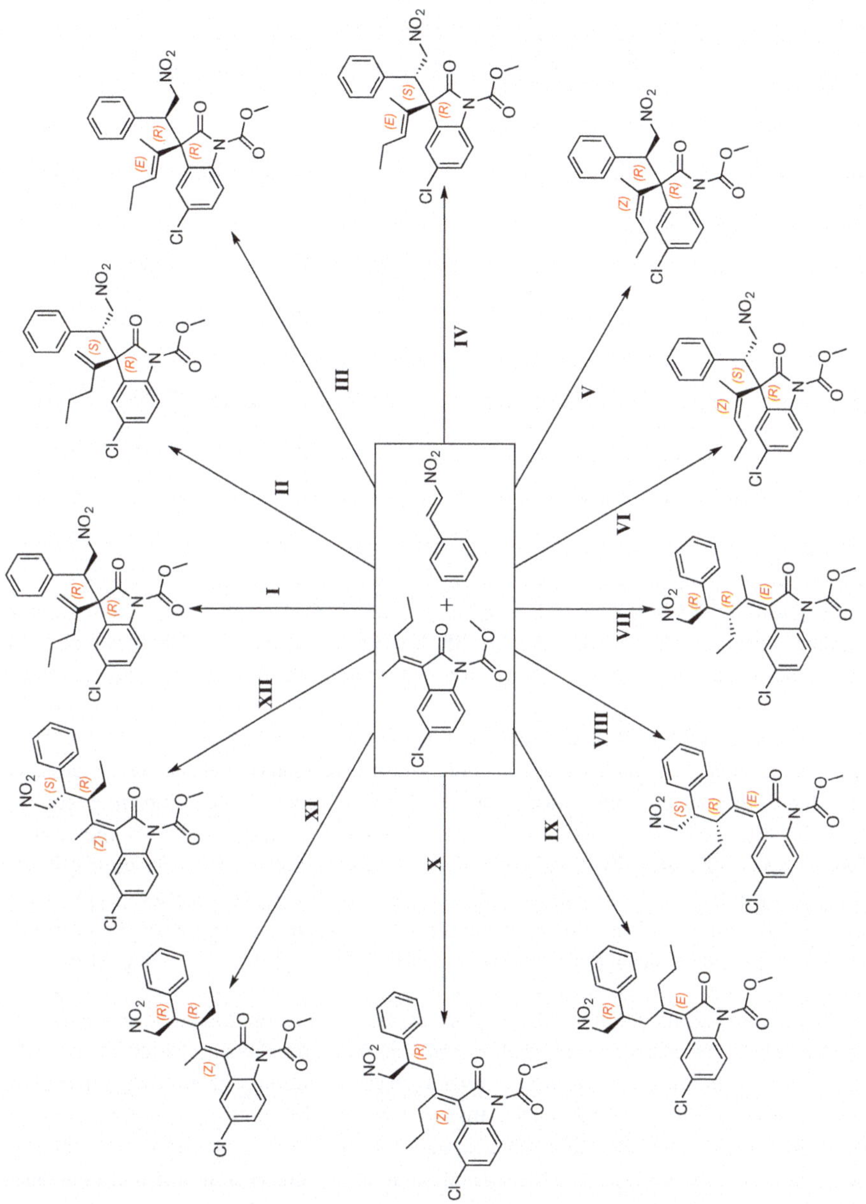

Fig. 2.4 All the possible products in their relative configurations

The first very important thing we noticed after the reactions is that we always obtained the product of γ-functionalization and we never observed the γ' adduct regardless of the starting oxindole. Next there is an evident KIE for the completely deuterated substrates with a 60% drop of reactivity for the **E** isomer (compare the

Table 2.1 Temperature screening for oxindole 84 (**a**) and 85 (**b**)

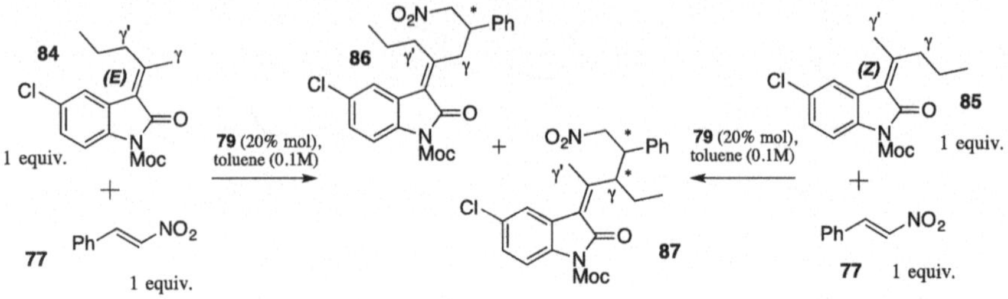

A) E-OXINDOLE **84**

Entry	Temp. (°C)	Yield (%) 86+87	86 : 87	ee% (86)	ee% (87)
1	r.t.	60	63 : 37	99	99
2	0	75	79 : 21	>99	>99
3	-20	85	100 : 0	>99	-

B) Z-OXINDOLE **85**

Entry	Temp. (°C)	Yield (%) 86+87	86 : 87	ee% (86)	ee% (87)
1	r.t.	67	14 : 86	99	99
2	0	80	5 : 95	-	>99
3	-20	86	0 : 100	-	>99

conversions of **84** and **89**) and an even greater 80% drop of reactivity for the **Z** isomer (compare the conversions of **85** and **93**). Finally, when we analyzed the reactions where only the methyl group is deuterated, we saw that for the **Z** isomer, having hydrogen atoms in γ, there was a nearly identical conversion (compare the conversions of **85** and **92**), whereas for the **E** isomer, having deuterium atoms in γ, there was a 20% drop in reactivity (compare the conversions of **84** and **88**). Hence, we concluded that the deprotonation is the RDS of the reaction, but more importantly, that the high selectivity does not come from the thermodynamics of the double bonds but comes from the catalyst which selectively and exclusively deprotonates the γ position. The discrimination between the two vinylogous positions is probably due also to some kind of hydrogen interaction between the oxygen of the carbonyl and the γ-protons that makes them more acidic with respect to the γ′ ones (Fig. 2.5).

With these results in our hands, we continued our investigation because we wanted to understand the exact role of reagents and catalyst and if we consider its structure there are two main units that are the tertiary base, eventually promoting a dienolate mechanism, and the H–bond-donor thiourea group, eventually promoting a dienol mechanism (Scheme 2.2). In order to understand how these functionalities interact with the reagents we “split” bifunctional catalyst **79** into two monofunctional ones **96** and **97** possessing only the tertiary base and the thiourea group respectively.

We used the **Z**-oxindole **85** as a probe reagent and left it under reaction conditions without nitrostyrene with each catalyst separately. We saw that every catalyst interacts with **85** and promotes its isomerization, particularly catalyst **97**. Unfortunately, from these results we still could not discriminate between the two mechanisms, so we

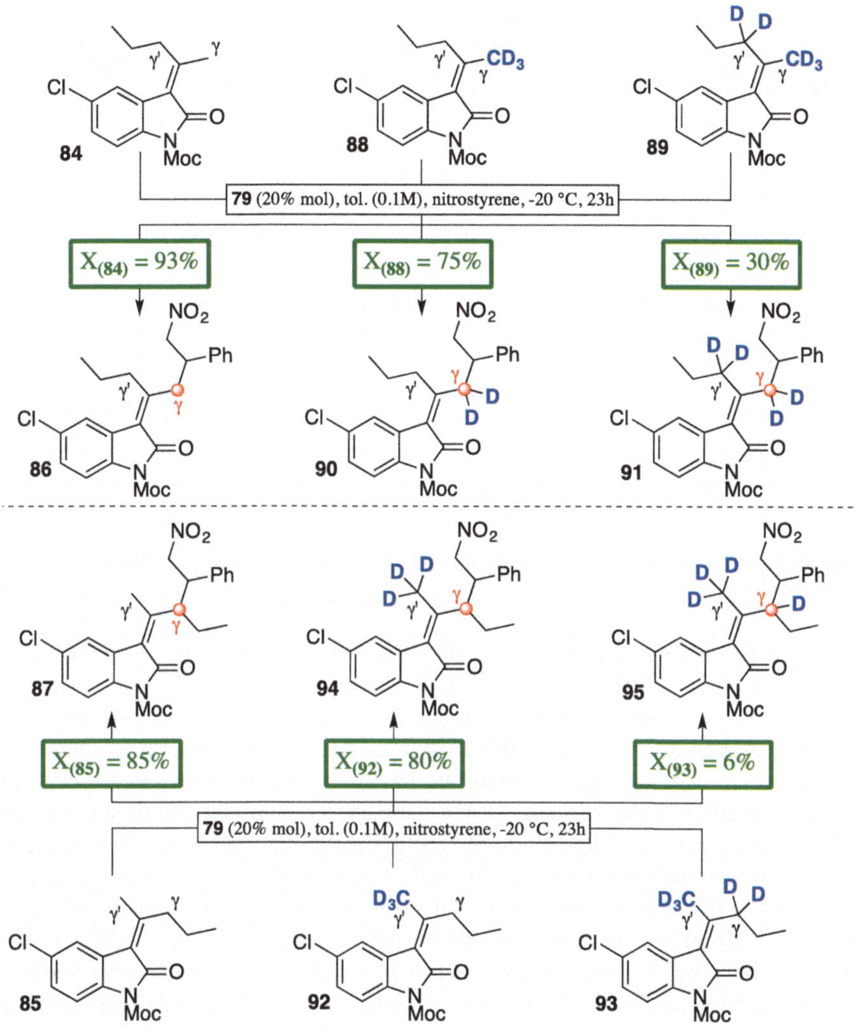

Scheme 2.1 Origin of the selectivity investigated with KIE

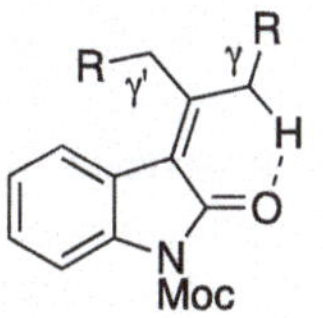

Fig. 2.5 Possible hydrogen interaction between the carbonyl and the γ protons

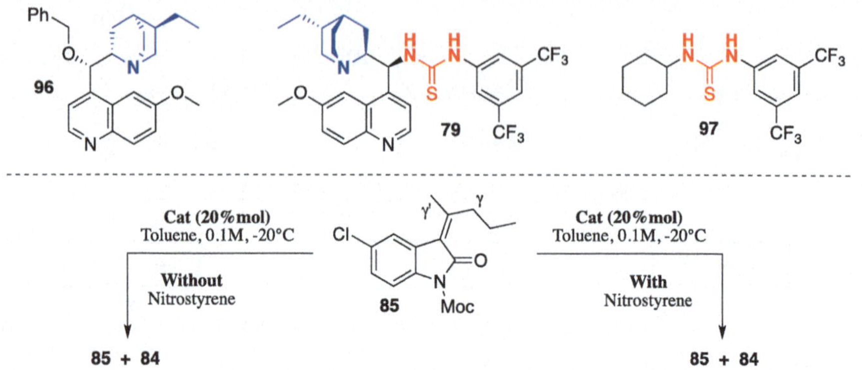

Without Nitrostyrene:

Entry	Catalyst	85:84 1h	3h	5h	23h
1	**79**	96:4	92:8	89:11	75:25
2	**96**	99:1	97:3	95:5	80:20
3	**97**	89:11	80:20	74:26	66:34

With Nitrostyrene:

Entry	Catalyst	85:84 1h	3h	5h	23h
1	**79**	-	-	-	-
2	**96**	99:1	98:2	97:3	80:20
3	**97**	99:1	98:2	97:3	96:4

Scheme 2.2 Interactions between the reagents and the functionalities of the catalyst

decided to repeat these experiments in the presence of nitrostyrene.[4] Even after 23 h, we did not observe formation of products whatsoever confirming what Casiraghi said about the necessity of having a bifunctional catalyst for this reaction. However, what is really important is that, with nitrostyrene, basic catalyst **96** still promotes the isomerization of **85** with an identical rate with respect to the previous experiment (compare entry 2 of the two tables). On the contrary with H–bond-donor catalyst **97**, the isomerization is almost completely stopped in the presence of nitrostyrene whereas in the previous case it was the fastest (compare entry 3 of the two tables). This is a very strong result suggesting that, during the reaction, the thiourea group of **79** interacts strongly and exclusively with the nitrostyrene, hence the oxindole is most likely activated by the base and the reaction proceeds via a dienolate intermediate. To confirm these results, we made two concentration experiments and keeping **85** as a probe, we repeated the reaction twice using two equivalents of oxindole in the first case and two equivalents of nitrostyrene in the second one (Table 2.2).

The reaction is much faster with a higher oxindole concentration giving full conversion after just 20 h and although we only obtained the reported product, we found the excess oxindole isomerized (entry 1). On the contrary, a higher concentration of nitrostyrene slows the reaction down without isomerization of the unreacted oxindole (entry 2). Summing all this information up we could confirm all our previous hypotheses that the nitrostyrene is activated by the thiourea of the catalyst forming a coordination compound that inhibits the isomerization of the oxindole which after selective deprotonation at the γ position (RDS) attacks the electrophile faster than it

[4]Naturally we did not repeat the experiment with catalyst **79** because it would promote the formation of the products and give us no information whatsoever on the reaction mechanism.

Table 2.2 Effects of concentration

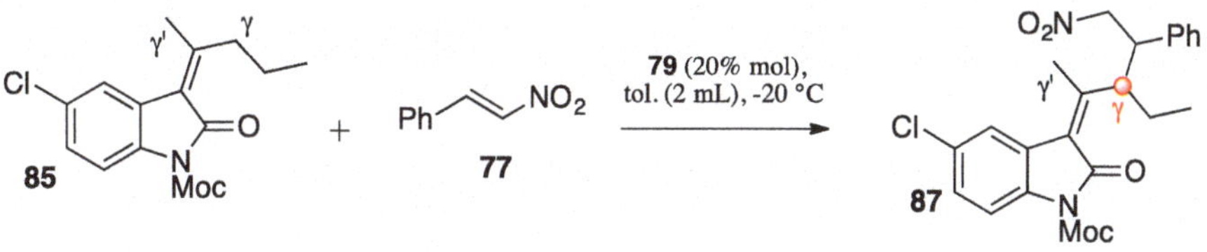

Entry	mmol 85	mmol 77	Time (h)	Conv. (%)	85:84
1	0.4	0.2	20	full	80:20
2	0.2	0.4	23	69	>99:1

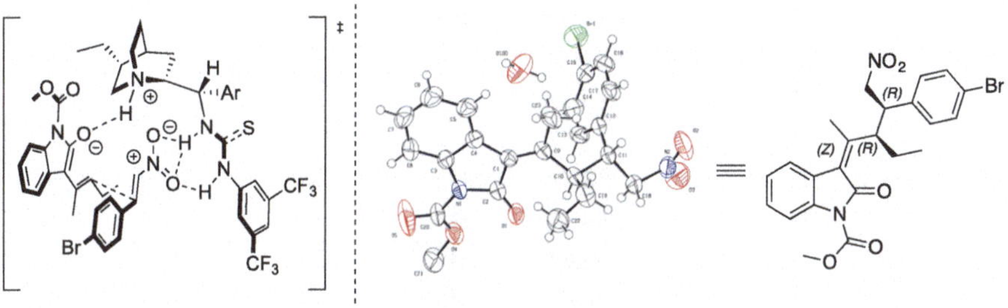

Fig. 2.6 Plausible transition state of the reaction and XRD structure

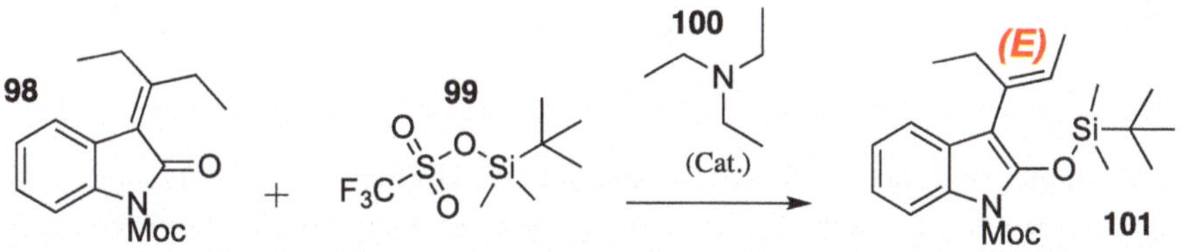

Reaction 2.2 Formation of the silyl ether

isomerizes. From single crystal XRD analysis we knew the absolute configuration of the products (*R,R*) so we could also draw a plausible TS that accounts for the selectivity of the reaction (Fig. 2.6).

The nitrostyrene interacts with the thiourea while the oxindole is activated by the base showing the *Si* face to the electrophile with the terminal double bond of the dienolate in the **E** configuration. One last experiment confirmed our assumption on the **E** configuration of the double bond in the TS (Reaction 2.2).

Generating the dienolate with TEA, which is a completely unselective base, and "trapping" it with TBDMSOTf showed that the double bond of the silyl dienol ether is formed exclusively in the **E** configuration.

At this point we moved forward to investigate the scope of the reaction starting from the **E** oxindole (Table 2.3).

Table 2.3 Scope of the E-oxindole

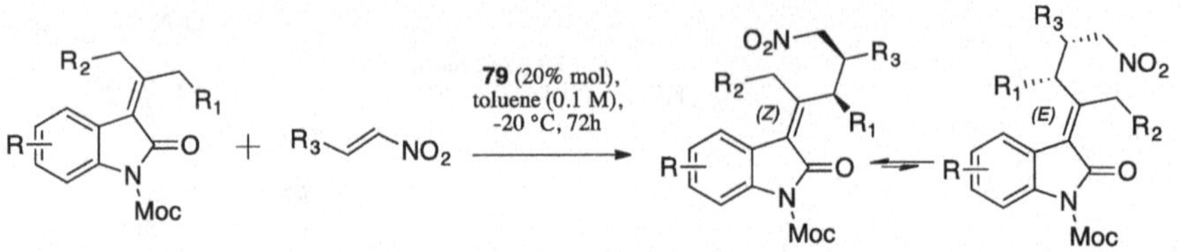

Entry	R	R1	R2	R3	Product	Yield (Z+E)	Z:E	ee
1	5-Cl	H	Ethyl	Ph	**106**	85%	91:9	>99%
2	5-Cl	H	Ethyl	Thiophenyl	**107**	77%	92:8	>99%
3	5-Cl	H	Ethyl	4-MeOPh	**108**	78%	94:6	>99%
4	5-Cl	H	Ethyl	3-MeOPh	**109**	92%	95:5	>99%
5	5-Cl	H	Ethyl	2-BnOPh	**110**	93%	93:7	>99%
6	5-Cl	H	Ethyl	4-MePh	**111**	96%	94:6	>99%
7	5-Cl	H	Ethyl	2-FPh	**112**	87%	95:5	>99%
8	5-Cl	H	Ethyl	4-BrPh	**113**	85%	>99:1	>99%
9	5-Cl	H	Ethyl	2,6-ClPh	**114**	80%	92:8	>99%
10	5-Cl	H	Ethyl	*i*-butyl	**115**	50%	>99:1	97%
11	6-Cl	H	Ethyl	Ph	**116**	75%	92:8	>99%
12	H	Methyl	Ethyl	Ph	**117**	44%	>99:1	>99%
13	H	H	*i*-butenyl	Ph	**118**	92%	80:20	>99%
14	H	H	Benzyl	Ph	**119**	63%	80:20	>99%

We immediately noticed that the products derived from the E-oxindole show some degree of isomerization but this time we knew it is a matter of thermodynamics of the double bond because we have already excluded an attack from the *s-trans* enolate from the previous experiments. To be even more sure we isolated the **Z** isomer and kept it under reaction conditions. After 72 h, we saw once again the same mixture of product that we obtained at the end of the reaction and concluded that this specific double bond is simply unstable in a basic environment. Aside from this phenomenon, the reaction of **E**-oxindoles is quite tolerant towards a variety of functional groups and in nearly all cases affords products with complete enantioselectivity so we moved on to investigate the **Z**-isomers that afford products with an additional stereocenter (Table 2.4).

We observed a trend very similar to the previous one with almost complete regio-diastereo- and enantioselectivity albeit with slightly lower yields most likely due to the greater steric hindrance provided by a generic alkyl substituent with respect to a methyl group. Finally, we were pleased to see that the reaction proceeds smoothly also when quaternary stereocenters are generated in the products (Scheme 2.3).

Table 2.4 Scope of the Z-oxindoles

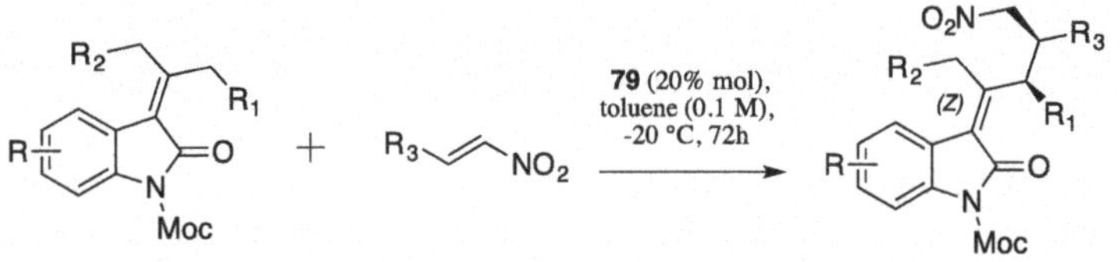

Entry	R	R1	R2	R3	Product	Yield	d.r.	ee
1	5-Cl	Ethyl	H	Ph	**120**	86%	>99:1	>99%
2	5-Cl	Ethyl	H	Thiophenyl	**121**	89%	>99:1	99%
3	H	Ethyl	H	Ph	**122**	70%	>99:1	>99%
4	H	Ethyl	H	4-MeOPh	**123**	52%	>99:1	98%
5	H	Ethyl	H	3-MeOPh	**124**	61%	>99:1	>99%
6	H	Ethyl	H	4-MePh	**125**	75%	>99:1	>99%
7	H	Ethyl	H	3-FPh	**126**	66%	>99:1	>99%
8	H	Ethyl	H	4-BrPh	**127**	62%	>99:1	>99%
9	H	Ethyl	H	*i*-butyl	**128**	20%	>99:1	99%
10	H	Ethyl	Methyl	Ph	**129**	44%	>99:1	99%
11	H	*i*-butenyl	H	Ph	**130**	81%	>99:1	>99%
12	H	Benzyl	H	Ph	**131**	50%	>99:1	96%

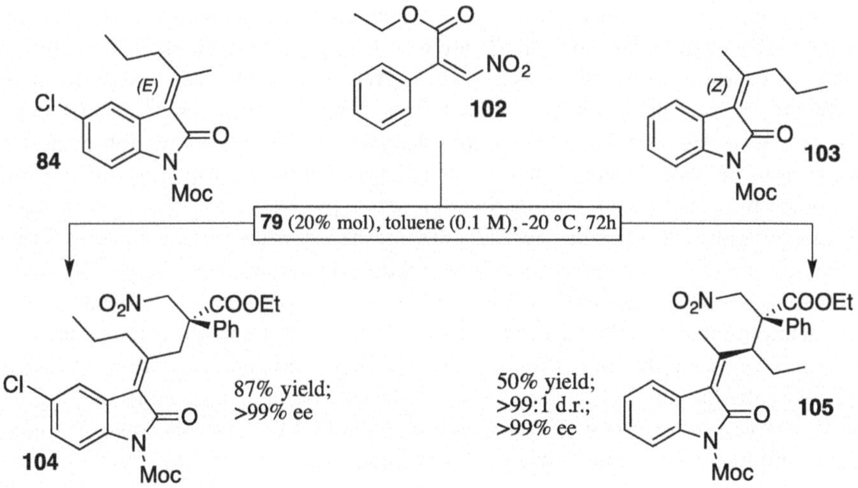

Scheme 2.3 Synthesis of the quaternary stereocenter derivatives

2.3 Conclusions

In conclusion, we have developed an organocatalyzed strategy for the vinylogous Michael addition of non-symmetric 3-alkylidene oxindoles to nitroolefin obtaining products with high regio-, diastereo- and enantiocontrol. The reaction proceeded exclusively via a selective deprotonation of the γ position of the oxindole by catalyst **79** which interacts via hydrogen bonding only with the nitroolefin. The role of the nitroalkene is furthermore fundamental because the direct interaction with the catalyst via H–bonding leads to a complex between the two species and reinforces the effect of the temperature in the inhibition of the isomerization of the oxindole double bond.

2.4 Experimental Section

2.4.1 General Information

The ^{1}H and ^{13}C NMR spectra were recorded at 400 and 100 MHz, respectively, or at 600 MHz for ^{1}H and 150 MHz for ^{13}C. All the ^{1}H and ^{13}C signals were assigned by means of g-COSY, g-HSQC and g-HMBC 2D-NMR sequences. NOE spectra were recorded using the DPFGSE-NOE sequence, using a mixing time of 1.0–2.0 s and "rsnob" 20 ÷ 50 Hz wide selective pulses, depending on the crowding of the spectra region. The chemical shifts (δ) for ^{1}H are given in ppm relative to the signals of internal standard TMS and for ^{13}C are given in ppm relative to the signals of the solvents. Coupling constants are given in Hz. When 2D-NMR were not performed, carbon types were determined from DEPT ^{13}C NMR experiments. ^{19}F NMR spectra were recorded with complete proton decoupling. The following abbreviations are used to indicate the multiplicity: s, singlet; d, doublet; t, triplet; q, quartet; m, multiplet; bs, broad signal. Purification of reaction products was carried out by flash chromatography (FC) on silica gel (230–400 mesh) according to the method of Still.[5] Organic solutions were concentrated under reduced pressure on a rotary evaporator. Optical rotations are reported as follows: $[\alpha]_D^{20}$ (*c* in g per 100 mL, solvent, % ee). Chiral thiourea catalyst **79** derived from 9-*epi*-9-amino-9-deoxy-dihydroquinine was prepared following the literature procedure.[6] Alkylidenoxindoles were synthesized following the literature procedure.[7] The diastereomeric ratio was determined by ^{1}H NMR analysis of the crude reaction mixture. Chiral HPLC analysis was performed using Amylose 2, Cellulose 2, AD-H, AS-H columns and OD-H with i-PrOH/hexane as the eluent were used.

---

[5]Still et al. [7].

[6]Vakulya et al. [8].

[7]Trost et al. [9].

2.4.2 Preparation of Deuterated Substrates

Methyl (*Z*)-5-chloro-2-oxo-3-(pentan-2-ylidene-1,1,1,3,3-*d5*)indoline-1-carboxylate and **methyl (*E*)-5-chloro-2-oxo-3-(pentan-2-ylidene-1,1,1,3,3-*d5*)indoline-1-carboxylate**

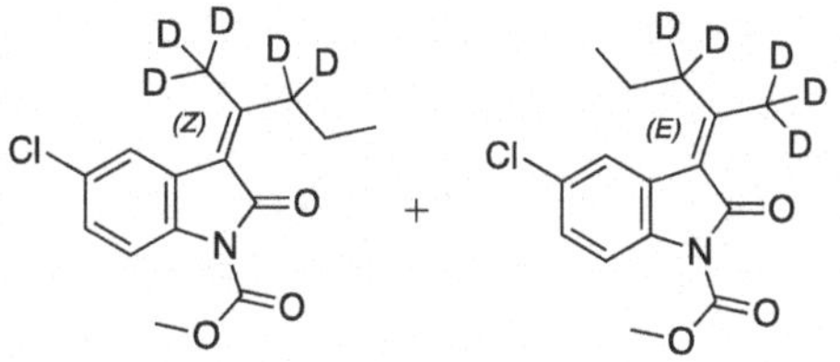

DABCO (0.225 mmol, 25 mg) was added to a 60:40 mixture of oxindoles (***Z***) and (***E***) (0.5 mmol, 150 mg) in the minimum amount of deuterated chloroform, then an excess of methanol-d_4 (2 mL) was added to the solution which was left under magnetic stirring at 40 °C until ^{1}H NMR confirmed the complete deuteration of the γ and γ' positions. At this point the solvent was removed at the rotary evaporator and the isomers were separated from each other by flash column chromatography (hexane/ethyl acetate = 9/1). MS-ESI (+): $(\boldsymbol{Z})_{d5}$ 321 [M+Na]$^+$; $(\boldsymbol{E})_{d5}$ 321 [M+Na]$^+$. ^{1}H NMR of $(\boldsymbol{Z})_{d5}$ (400 MHz, $CDCl_3$): δ (ppm): 7.93 (*d*, 1H, J_1 = 9.0 Hz); 7.52 (*d*, 1H, J_1 = 2.0 Hz); 7.25 (*dd*, 1H, J_1 = 9.0 Hz, J_2 = 2.0 Hz); 4.02 (*s*, 3H); 1.59 (*m*, 2H); 1.03 (*t*, 3H, J_1 = 7.5 Hz). ^{13}C NMR: δ (ppm): 164.3, 164.2, 151.7, 135.9, 129.4, 127.5, 125.7, 123.2, 120.5, 115.7, 53.8, 21.3, 14.2. ^{1}H NMR of $(\boldsymbol{E})_{d5}$ (400 MHz, $CDCl_3$): δ (ppm): 7.95 (*d*, 1H, J_1 = 9.2 Hz); 7.45 (*d*, 1H, J_1 = 2.1 Hz); 7.26 (*dd*, 1H, J_1 = 9.2 Hz, J_2 = 2.1 Hz); 4.02 (*s*, 3H); 1.67 (*m*, 2H); 1.11 (*t*, 3H, J_1 = 7.3 Hz). ^{13}C NMR: δ (ppm): 165.1, 164.3, 151.7, 136.0, 129.5, 127.6, 124.9, 122.8, 120.3, 115.8, 53.8, 20.2, 14.2.

Methyl (*Z*)-5-chloro-2-oxo-3-(pentan-2-ylidene-1,1,1-*d3*)indoline-1-carboxylate and **methyl (*E*)-5-chloro-2-oxo-3-(pentan-2-ylidene-1,1,1-*d3*)indoline-1-carboxylate**

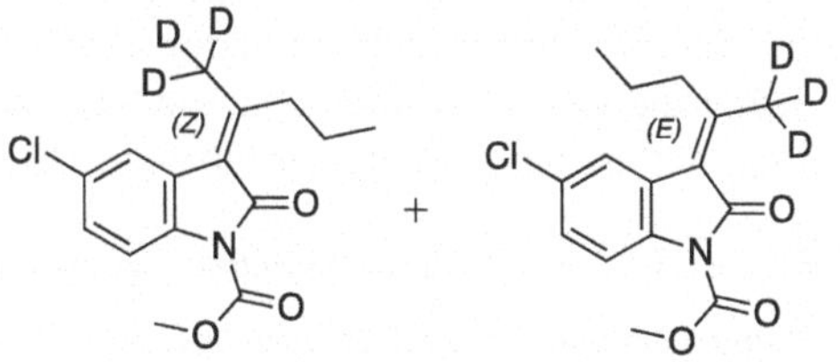

DABCO (0.53 mmol, 59.0 mg) was added to was added to a 60:40 mixture of oxindoles (***Z***) and (***E***) (1.2 mmol, 356 mg) in 7 mL of deuterated chloroform, then methanol-d_4 (4.2 equiv., 170 μL, 150 mg) was added to the solution which was left under magnetic stirring overnight at room temperature. At this point the solvent was removed at the rotary evaporator and the isomers were separated from each other by flash column chromatography (hexane/ethyl acetate = 9/1). From the ^{1}H NMR spectra we calculated the deuterium enrichment for each position of each isomer. Both isomers have a 50% deuterium enrichment on the methyl group and a 10% enrichment on the propyl group. MS-ESI (+): $(\mathbf{Z})_{d3}$ 319 [M+Na]$^+$; $(\boldsymbol{E})_{d3}$ 319 [M+Na]$^+$. ^{1}H NMR of $(\mathbf{Z})_{d3}$ (400 MHz, $CDCl_3$): δ (ppm): 7.91 (*d*, 1H, $J_1 = 9.4$ Hz); 7.51 (*d*, 1H, $J_1 = 2.2$ Hz); 7.23 (*dd*, 1H, $J_1 = 9.4$ Hz, $J_2 = 2.2$ Hz); 4.02 (*s*, 3H); 3.00 (*m*, 1.8H); 2.36 (*m*, 1.45H); 1.60 (*m*, 2H); 1.04 (*t*, 3H, $J_1 = 7.4$ Hz). ^{13}C NMR: δ (ppm): 164.3, 164.2, 151.6, 135.8, 129.4, 127.5, 125.6, 123.2, 120.4, 115.7, 53.8, 38.7, 24.3, 21.3, 14.3. ^{1}H NMR of $(\boldsymbol{E})_{d3}$ (400 MHz, $CDCl_3$): δ (ppm): 7.93 (*d*, 1H, $J_1 = 8.8$ Hz); 7.42 (*d*, 1H, $J_1 = 2.2$ Hz); 7.24 (*dd*, 1H, $J_1 = 8.8$ Hz, $J_2 = 2.821$ Hz); 4.02 (*s*, 3H); 2.63 (*m*, 1.78H); 2.54 (*m*, 1.51H); 1.67 (*m*, 2H); 1.11 (*t*, 3H, $J_1 = 7.7$ Hz). ^{13}C NMR: δ (ppm): 165.0, 164.4, 151.6, 135.9, 129.5, 127.6, 124.8, 122.8, 120.2, 115.8, 53.8, 40.4, 22.6, 20.3, 14.3.

2.4.3 General Procedure for the Preparation of Alkylidenoxindoles

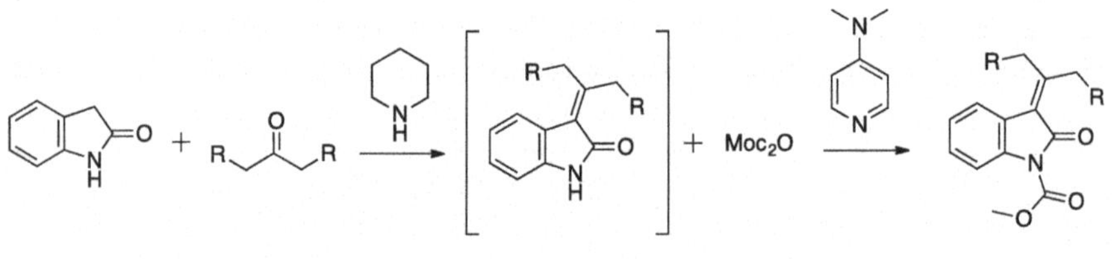

Piperidine (4 equiv.) was added to a 0.5 M solution of oxindole (1 equiv.) in ethanol:ketone 1:1. The resulting solution was stirred overnight at room temperature. The reaction mixture was taken up with ethyl acetate and the resulting organic solution was respectively washed with 20 mL of 1 M solution of $KHSO_4$, water and brine. The organic layer was made anhydrous over $MgSO_4$, filtered and evaporated under reduced pressure. The crude residue was suspended in acetonitrile and DMAP (0.1 equiv.) was added followed by the addition of dimethyl dicarbonate (1.2 equiv.). After 30 min of stirring the solvent was removed under reduced pressure and the crude products was purified by flash column chromatography.

(*E*)-methyl-5-chloro-2-oxo-3-(pentan-2-ylidene)indoline-1-carboxylate and **(*Z*)-methyl-5-chloro-2-oxo-3-(pentan-2-ylidene)indoline-1-carboxylate**

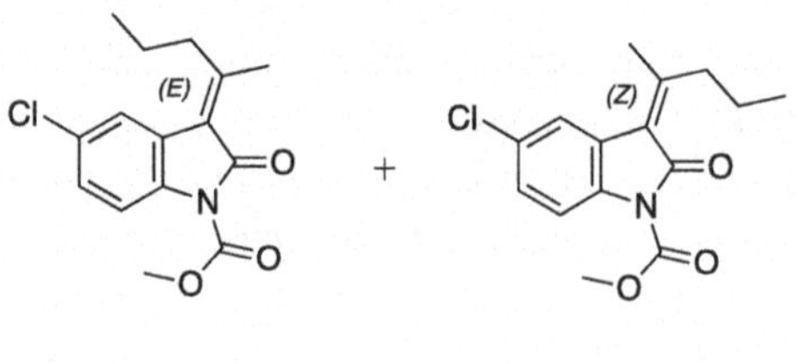

The title compounds were obtained following the general procedure as a 45:55 mixture of **E**:**Z** stereoisomers. These were purified and separated from each other by flash column chromatography (hexane/ethyl acetate = 9/1) to give an overall yield of 75%. HRMS-ESI (+) of (***E***): calculated for $C_{15}H_{16}ClNaNO_3$ 316.0716, found 316.0718 $[M+Na]^+$, and of (**Z**) calculated for $C_{15}H_{16}ClNaNO_3$ 316.0716, found 316.0717 $[M+Na]^+$. ^{1}H NMR of (***E***) (400 MHz, $CDCl_3$): δ (ppm): 7.93 (*d*, 1H, J = 8.6 Hz); 7.42 (*d*, 1H, J = 2.7 Hz); 7.24 (*dd*, 1H, J_1 = 8.6 Hz, J_2 = 2.7 Hz); 4.02 (*s*, 3H); 2.63 (*t*, 2H, J = 8.4 Hz); 2.57 (*s*, 3H); 1.68 (*m*, 2H); 1.11 (*t*, 3H, J = 7.3 Hz). ^{13}C NMR: δ (ppm): 165.0, 164.4, 151.7, 135.9, 129.5, 127.6, 124.8, 122.8, 120.2, 115.8, 53.8, 40.5, 22.9, 20.4, 14.3. ^{1}H NMR of (**Z**) (400 MHz, $CDCl_3$): δ (ppm): 7.93 (*d*, 1H, J = 8.5 Hz); 7.53 (*d*, 1H, J_1 = 1.7 Hz); 7.25 (*dd*, 1H, J_1 = 8.5 Hz, J_2 = 1.6 Hz); 4.03 (*s*, 3H); 3.02 (*t*, 2H, J = 7.8 Hz); 2.38 (*s*, 3H); 1.59 (*m*, 2H); 1.04 (*t*, 3H, J = 7.4 Hz). ^{13}C NMR (100 MHz, $CDCl_3$): δ (ppm): 164.4, 151.7, 135.9, 129.5, 127.6, 125.7, 123.3, 120.5, 115.7, 53.9, 38.9, 24.6, 21.5, 14.3.

(*Z*)-methyl-2-oxo-3-(pentan-2-ylidene)indoline-1-carboxylate

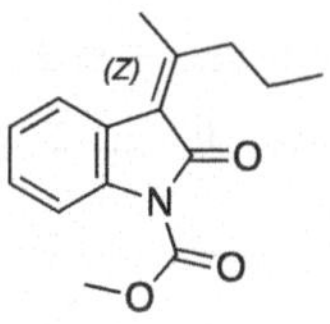

The title compound was obtained following the general procedure in 45% yield after purification of the crude mixture by flash column chromatography (hexane/ethyl acetate = 9/1). HRMS-ESI (+): calculated for $C_{15}H_{17}NaNO_3$ 282.1106, found 282.1104 $[M+Na]^+$. ^{1}H NMR (400 MHz, $CDCl_3$): δ (ppm): 7.88 (*d*, 1H, J = 7.8 Hz); 7.43 (*d*, 1H, J = 7.7 Hz); 7.19 (*m*, 1H); 7.06 (*m*, 1H); 3.98 (*s*, 3H); 2.95 (*t*, 2H, J = 8.2 Hz); 2.27(*s*, 3H); 1.56 (*m*, 2H); 1.02 (*t*, 3H, J = 7.2 Hz). ^{13}C NMR (100 MHz, $CDCl_3$): δ (ppm): 165.1, 162.3, 151.9, 137.6, 129.9, 127.9, 124.0, 123.3, 121.2, 114.7, 53.7, 38.7, 24.5, 21.5, 14.3.

(*E*)-methyl-3-(hexan-3-ylidene)-2-oxoindoline-1-carboxylate and **(*Z*)-methyl-3-(hexan-3-ylidene)-2-oxoindoline-1-carboxylate**

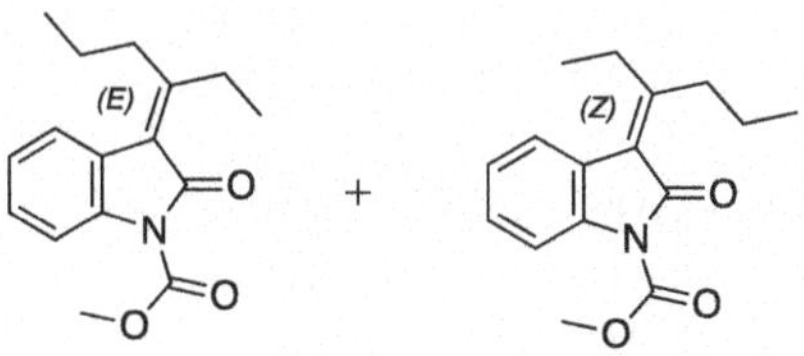

The title compounds were obtained following the general procedure as a 1:1 mixture of **E**:**Z** stereoisomers. These were purified by flash column chromatography (hexane/ethyl acetate = 9/1) to give an overall yield of 40% and separated from each other by preparative reverse-phase column chromatography. HRMS-ESI (+) of (***E***): calculated for $C_{16}H_{19}NaNO_3$ 296.1263, found 296.1261 $[M+Na]^+$ and of (**Z**) calculated for $C_{16}H_{19}NaNO_3$ 296.1263, found 296.1261 $[M+Na]^+$. ^1H NMR of (***E***) (400 MHz, $CDCl_3$): δ (ppm): 8.00 (*d*, 1H, *J* = 8.6 Hz); 7.47 (*d*, 1H, *J* = 8.4 Hz); 7.28 (*m*, 1H); 7.16 (*m*, 1H); 4.02 (*s*, 3H); 2.98 (*q*, 2H, *J* = 7.6 Hz); 2.63 (*t*, 2H, *J* = 8.5 Hz); 1.66 (*m*, 2H); 1.18 (*t*, 3H, *J* = 7.8 Hz); 1.11 (*t*, 3H, *J* = 7.5 Hz). ^{13}C NMR (100 MHz, $CDCl_3$): δ (ppm): 168.3, 165.2, 151.9, 137.7, 127.9, 124.1, 123.6, 122.8, 120.4, 114.7, 53.7, 38.5, 28.7, 20.4, 14.5, 12.4. ^1H NMR of (**Z**) (300 MHz, $CDCl_3$): δ (ppm): 8.01 (*d*, 1H, *J* = 9.0 Hz); 7.56 (*d*, 1H, *J* = 7.6 Hz); 7.30 (*m*, 1H); 7.18 (*m*, 1H); 4.03 (*s*, 3H); 2.94 (*t*, 2H, *J* = 8.1 Hz); 2.71 (*q*, 2H, *J* = 7.5 Hz); 2.38 (*s*, 3H); 1.59 (*m*, 2H); 1.26 (*t*, 3H, *J* = 7.6 Hz); 1.05 (*t*, 3H, *J* = 7.6 Hz). ^{13}C NMR (100 MHz, $CDCl_3$): δ (ppm): 168.2, 165.3, 151.9, 137.7, 127.9, 124.2, 123.5, 122.9, 120.5, 114.7, 53.7, 37.0, 29.8, 21.7, 14.6, 11.2.

(*E*)-methyl-3-(5-methylhex-5-en-2-ylidene)-2-oxoindoline-1-carboxylate and **(*Z*)-methyl-3-(5-methylhex-5-en-2-ylidene)-2-oxoindoline-1-carboxylate**

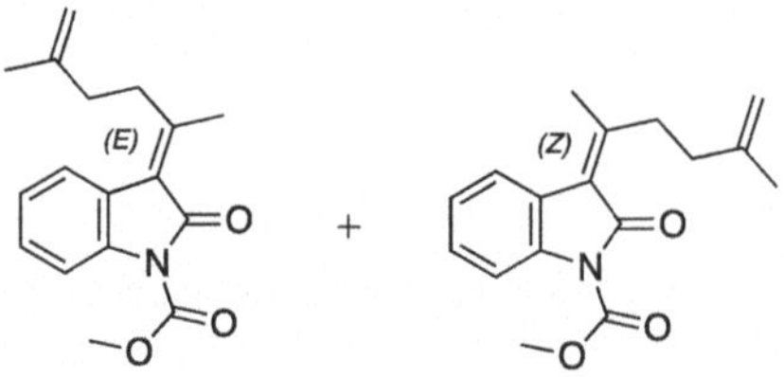

The title compounds were obtained following the general procedure as a 40:60 mixture of **E**:**Z** stereoisomers. These were purified and separated from each other by flash column chromatography (hexane/ethyl acetate = 9/1) to give an overall yield of 58%. HRMS-ESI (+) of (***E***): calculated for $C_{17}H_{19}NaNO_3$ 308.1263, found 308.1265 $[M+Na]^+$ and of (**Z**) calculated for $C_{17}H_{19}NaNO_3$ 308.1263, found 308.1262 $[M+Na]^+$. ^1H NMR of (***E***) (300 MHz, $CDCl_3$): δ (ppm): 8.00 (*d*, 1H, *J* = 8.5 Hz); 7.50 (*d*, 1H, *J* = 7.6 Hz); 7.29 (*m*, 1H); 7.16 (*m*, 1H); 4.84 (*d*, 2H, *J* = 11.6 Hz); 4.03 (*s*, 3H); 2.82 (*t*, 2H, *J* = 8.4 Hz); 2.57 (*s*, 3H); 2.29 (*t*, 2H, *J* = 8.4 Hz); 1.83 (*s*, 3H). ^{13}C NMR (100 MHz, $CDCl_3$): δ (ppm): 165.6, 161.5, 151.8, 144.2, 137.6, 128.0, 124.2, 123.3, 122.6, 121.1, 114.7, 110.8, 53.6, 36.8, 34.3, 22.6, 22.5. ^1H NMR of (**Z**) (300 MHz, $CDCl_3$): δ (ppm): 7.99 (*d*, 1H, *J* = 8.3 Hz); 7.59 (*d*, 1H, *J* = 7.2 Hz); 7.31 (*m*, 1H); 7.18 (*m*, 1H); 4.76 (*bs*, 2H); 4.03 (*s*, 3H); 3.19 (*t*, 2H, *J* = 8.1 Hz); 2.40 (*s*, 3H); 2.26 (*t*, 2H, *J* = 7.6 Hz); 1.83 (*s*, 3H). ^{13}C NMR (100 MHz, $CDCl_3$): δ (ppm): 164.9, 161.3, 151.8, 144.9, 137.6, 127.9, 124.2, 123.9, 123.3, 121.3, 114.6, 110.5, 53.7, 35.7, 35.3, 24.4, 22.3.

(*E*)-methyl-2-oxo-3-(4-phenylbutan-2-ylidene)indoline-1-carboxylate and **(*Z*)-methyl 2-oxo-3-(4-phenylbutan-2-ylidene)indoline-1-carboxylate**

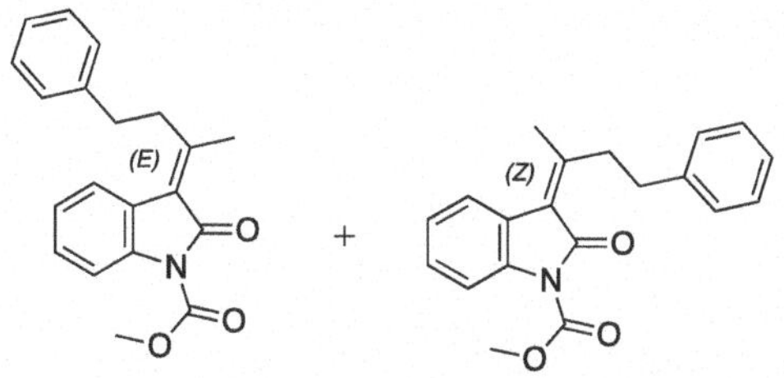

The title compounds were obtained following the general procedure as a 40:60 mixture of **E**:**Z** stereoisomers. These were purified and separated from each other by flash column chromatography (hexane/ethyl acetate = 9/1) to give an overall yield of 62%. HRMS-ESI (+) of (***E***): calculated for $C_{20}H_{19}NaNO_3$ 344.1263, found 344.1263 $[M+Na]^+$ and of (**Z**) calculated for $C_{20}H_{19}NaNO_3$ 344.1263, found 344.1266 $[M+Na]^+$. ^{1}H NMR of (***E***) (300 MHz, $CDCl_3$): δ (ppm): 8.03 (*d*, 1H, $J = 8.4$ Hz); 7.60 (*d*, 1H, $J = 7.5$ Hz); 7.35–7.20 (*m*, 7H); 4.04 (*s*, 3H); 3.00 (*m*, 4H); 2.55 (*s*, 3H). ^{13}C NMR (100 MHz, $CDCl_3$): δ (ppm): 165.7, 160.8, 151.9, 140.4, 137.7, 128.7, 128.3, 128.2, 128.1, 126.5, 124.2, 123.3.122.7, 121.4, 114.9, 53.8, 40.3, 32.8, 22.9. ^{1}H NMR of (**Z**) (300 MHz, $CDCl_3$): δ (ppm): 8.01 (*d*, 1H, $J = 8.0$ Hz); 7.59 (*d*, 1H, $J_1 = 8.00$ Hz); 7.31 (*m*, 5H); 7.19 (*m*, 2H); 4.05 (*s*, 3H); 3.32 (*t*, 2H, $J = 8.2$ Hz); 2.87 (*q*, 2H, $J = 8.2$ Hz); 2.36 (*s*, 3H). ^{13}C NMR (100 MHz, $CDCl_3$): δ (ppm): 164.8, 160.7, 151.8, 141.2, 137.7, 128.5, 128.3, 128.0, 126.0, 124.1, 124.0, 123.4, 121.6, 114.7, 53.7, 39.3, 34.2, 24.8.

(*E*)-methyl 6-chloro-2-oxo-3-(pentan-2-ylidene)indoline-1-carboxylate

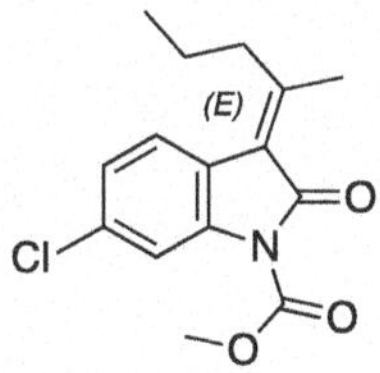

The title compound was obtained following the general procedure in 51% yield after purification of the crude mixture by flash column chromatography (hexane/ethyl acetate = 9/1). HRMS-ESI (+): calculated for $C_{15}H_{16}ClNaNO_3$ 316.0716, found 316.0717 $[M+Na]^+$. ^{1}H NMR (400 MHz, $CDCl_3$): δ (ppm): 8.05 (*d*, 1H, $J = 2.1$ Hz); 7.40 (*d*, 1H, $J = 8.5$ Hz); 7.15 (*dd*, 1H, $J_1 = 8.5$ Hz, $J_2 = 2.1$ Hz); 4.03 (*s*, 3H); 2.64 (*t*, 2H, $J = 8.3$ Hz); 2.56 (*s*, 3H); 1.67 (*m*, 2H); 1.10 (*t*, 3H, $J = 7.3$ Hz). ^{13}C NMR (100 MHz, $CDCl_3$): δ (ppm): 165.3, 163.2, 151.7, 138.3, 133.6, 124.2, 123.5, 121.9, 120.3, 115.3, 53.9, 40.6, 22.8, 20.3, 14.3.

2.4.4 General Procedure for the Vinylogous Michael Addition of Non-symmetric 3-Alkylidene Oxindoles to Nitroalkenes

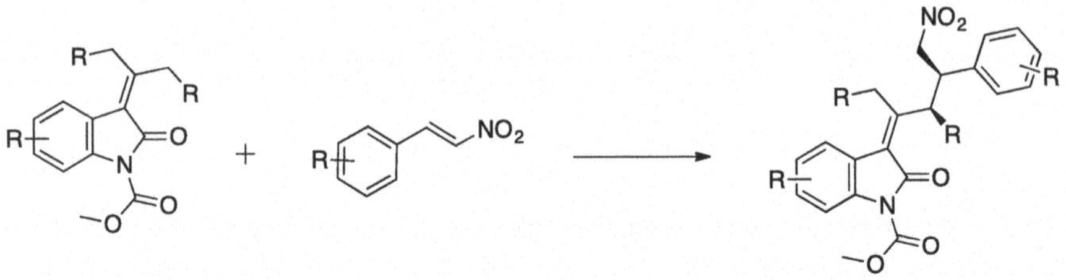

All the reaction were carried out in undistilled toluene. In an ordinary vial equipped with a Teflon-coated stir bar containing 3-alkylidenoxindole derivative (0.2 mmol, 1.0 equiv.), nitroalkene (0.2 mmol, 1.0 equiv.), 9-*epi*-9-amino-9-deoxy-dihydroquinine **79** (0.04 mmol, 0.2 equiv.) and 2 mL of toluene were added. The resulting solution was stirred at −20 °C for 72 h. The crude mixture was flushed through a short plug of silica, using dichloromethane/ethyl acetate 1:1 as the eluent (50 ml). Solvent was removed under reduced pressure and the diastereomeric ratio (dr) was determined by ^{1}H NMR analysis of the crude mixture. The desired compound was isolated by flash column chromatography.

(*R*,*Z*)-methyl-5-chloro-3-(1-nitro-2-phenylheptan-4-ylidene)-2-oxoindoline-1-carboxylate (product **106**, Table 2.3—entry 1)

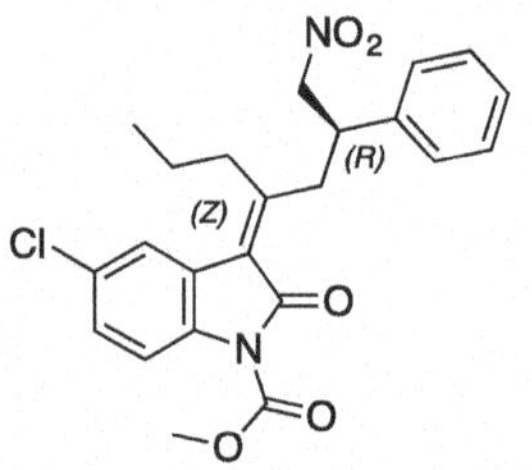

The reaction was carried out following the general procedure to furnish the crude product **106** as 91:9 mixture of (*Z*)-**106** and (*E*)-**106**. The crude mixture has been purified by flash column chromatography (hexane/ethyl acetate = 8/2) to give an overall yield of 85% and (*Z*)-**106** in a >99% ee. HPLC analysis on a OD-H column: hexane/*i*-PrOH 90/10, flow rate 0.75 mL/min, $\lambda = 214$ nm: $\tau_{major} = 20.79$ min. $[\alpha]_D^{20}$ −120.0 (*c* 1.00, $CHCl_3$). HRMS-ESI (+): calculated for $C_{23}H_{23}ClNaN_2O_5$ 465.1188, found 465.1188 $[M+Na]^+$. ^{1}H NMR (600 MHz, $CDCl_3$): δ (ppm): 7.95 (*d*, 1H, $J = 8.8$ Hz); 7.40 (*d*, 1H, $J = 2.0$ Hz); 7.30 (*m*, 6H); 4.79 (*dd*, 1H, $J_1 = 13.1$ Hz, $J_2 = 9.9$ Hz); 4.71 (*dd*, 1H, $J_1 = 12.9$ Hz, $J_2 = 5.6$ Hz); 4.07 (*s*, 3H); 4.03 (*dd*, 1H, $J_1 = 12.2$ Hz, $J_2 = 8.2$ Hz); 3.85 (*m*, 1H); 2.83 (*dd*, 1H, $J_1 = 12.1$ Hz,

$J_2 = 7.4$ Hz); 2.64 (*m*, 1H); 2.23 (*m*, 1H); 1.54 (*m*, 2H); 1.05 (*t*, 3H, $J = 7.3$ Hz). ^{13}C NMR (150 MHz, $CDCl_3$): δ (ppm): 165.1 (C); 163.3 (C); 151.4 (C); 139.2 (C); 136.3 (C); 129.9 (C); 129.0 (CH); 128.5 (CH); 128.0 (CH); 127.5 (CH); 124.1 (C); 123.4 (CH); 122.6 (C); 116.0 (CH); 79.4 (CH_2); 54.1 (CH_3); 43.7 (CH); 38.7 (CH_2); 38.4 (CH_2); 20.3 (CH_2); 14.3 (CH_3).

(*S*,*Z*)-methyl-5-chloro-3-(1-nitro-2-(thiophen-2-yl)heptan-4-ylidene)-2-oxoindoline-1-carboxylate (product **107**, Table 2.3—entry 2)

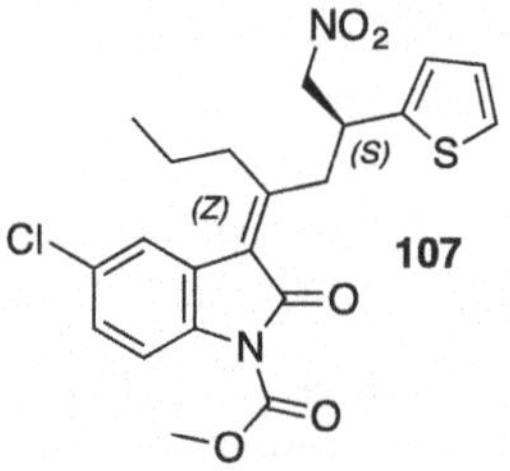

107

The reaction was carried out following the general procedure to furnish the crude product **107** as 92:8 mixture of (*Z*)-**107** and (*E*)-**107**. The crude mixture has been purified by flash column chromatography (hexane/ethyl acetate = 8/2) to give an overall yield of 77% and (*Z*)-**107** in a >99% ee. HPLC analysis on a AD-H column: hexane/*i*-PrOH 95/5, flow rate 1.0 mL/min, $\lambda = 254$ nm: $\tau_{major} = 14.82$ min. $[\alpha]_D^{20}$ −344.4 (*c* 1.00, $CHCl_3$). HRMS-ESI (+): calculated for $C_{21}H_{21}ClNaN_2O_5S$ 471.0752, found 471.0750 $[M+Na]^+$. ^{1}H NMR (600 MHz, $CDCl_3$): δ (ppm): 7.96 (*d*, 1H, $J = 8.9$ Hz); 7.42 (*d*, 1H, $J = 2.0$ Hz); 7.31 (*dd*, 1H, $J_1 = 8.9$ Hz, $J_2 = 1.9$ Hz); 7.22 (*dd*, 1H, $J_1 = 4.6$ Hz, $J_2 = 1.5$ Hz); 6.94 (*m*, 2H); 4.75 (*m*, 2H); 4.18 (*m*, 1H); 4.06 (*s*, 3H); 3.99 (*m*, 1H); 2.90 (*dd*, 1H, $J_1 = 12.0$ Hz, $J_2 = 7.9$ Hz); 2.67 (*m*, 1H); 2.23 (*m*, 1H); 1.55 (*m*, 2H); 1.06 (*t*, 3H, $J = 7.3$ Hz). ^{13}C-NMR (150 MHz, $CDCl_3$): δ (ppm): 165.0, 162.9, 151.3, 141.9, 136.3, 129.9, 128.5, 127.1, 125.6, 124.8, 124.0, 123.5, 122.6, 116.1, 80.1, 54.1, 39.7, 39.0, 38.5, 20.1, 14.3.

(*R*,*Z*)-methyl-5-chloro-3-(2-(4-methoxyphenyl)-1-nitroheptan-4-ylidene)-2-oxoindoline-1-carboxylate (product **108**, Table 2.3—entry 3)

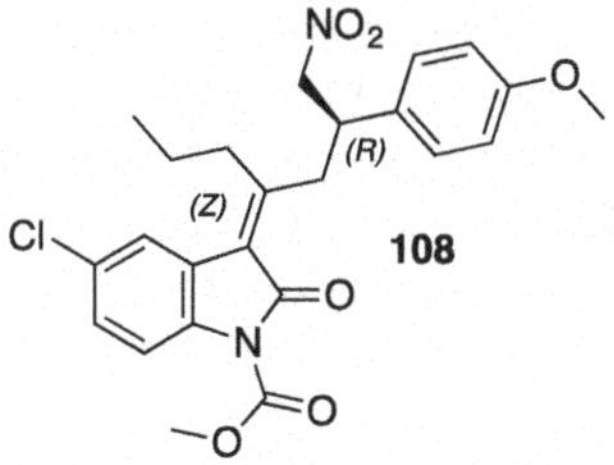

108

The reaction was carried out following the general procedure to furnish the crude product **108** as 96:4 mixture of (Z)-**108** and (*E*)-**108**. The crude mixture has been purified by flash column chromatography (hexane/ethyl acetate = 8/2) to give an overall yield of 78% and (Z)-**108** in a >99% ee. HPLC analysis on a AD-H column: hexane/*i*-PrOH 95/5, flow rate 1 mL/min, $\lambda = 214$ nm: $\tau_{major} = 17.38$. $[\alpha]_D^{20}$ −83.0 (*c* 1.00, $CHCl_3$). HRMS-ESI (+): calculated for $C_{24}H_{25}ClNaN_2O_6$ 495.1293, found 495.1292 $[M+Na]^+$. ^1H NMR (600 MHz, $CDCl_3$): δ (ppm): 7.95 (*d*, 1H, $J = 8.8$ Hz); 7.40 (*d*, 1H, $J = 2.1$ Hz); 7.30 (*dd*, 1H, $J_1 = 8.8$ Hz, $J_2 = 2.1$ Hz); 7.22 (*d*, 1H, $J = 8.6$ Hz); 6.85 (*d*, 1H, $J = 8.6$ Hz); 4.74 (*dd*, 1H, $J_1 = 12.6$ Hz, $J_2 = 9.9$ Hz); 4.68 (*dd*, 1H, $J_1 = 12.6$ Hz, $J_2 = 5.6$ Hz); 4.06 (*s*, 3H); 3.97 (*dd*, 1H, $J_1 = 12.2$ Hz, $J_2 = 7.9$ Hz); 3.80 (*m*, 4H); 2.84 (*dd*, 1H, $J_1 = 12.2$ Hz, $J_2 = 7.7$ Hz); 2.64 (*m*, 1H); 2.24 (*m*, 1H); 1.55 (*m*, 1H); 1.05 (*t*, 3H, $J = 7.4$ Hz). ^{13}C-NMR (150 MHz, $CDCl_3$): δ (ppm): 165.1, 163.6, 159.2, 151.4, 136.2, 131.0, 129.9, 128.5, 128.4, 124.1, 123.4, 122.5, 116.0, 114.3, 79.7, 55.2, 54.1, 43.0, 38.7, 38.4, 20.2, 14.3.

(*R*,*Z*)-methyl-5-chloro-3-(2-(3-methoxyphenyl)-1-nitroheptan-4-ylidene)-2-oxoindoline-1-carboxylate (product **109**, Table 2.3—entry 4)

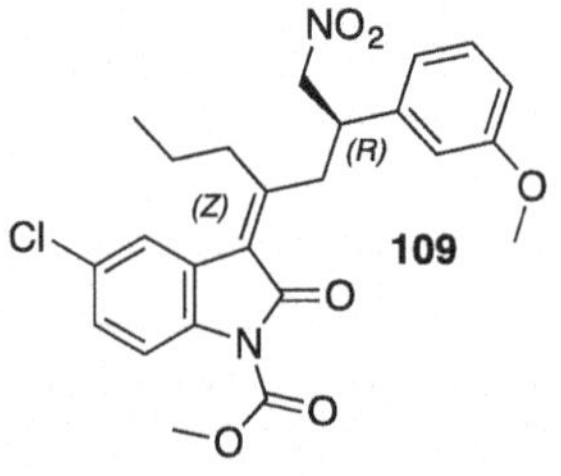

109

The reaction was carried out following the general procedure to furnish the crude product **109** as 95:5 mixture of (Z)-**109** and (*E*)-**109**. The crude mixture has been purified by flash column chromatography (hexane/ethyl acetate = 8/2) to give an overall yield of 92% and (Z)-**109** in a >99% ee. HPLC analysis on a AD-H column: hexane/*i*-PrOH 95/5, flow rate 1 mL/min, $\lambda = 214$ nm: $\tau_{major} = 14.04$ min. $[\alpha]_D^{20}$ −79.5 (*c* 1.00, $CHCl_3$). HRMS-ESI (+): calculated for $C_{24}H_{25}ClNaN_2O_6$ 495.1293, found 495.1292 $[M+Na]^+$. ^1H NMR (600 MHz, $CDCl_3$): δ (ppm): 7.95 (*d*, 1H, $J = 8.9$ Hz); 7.41 (*d*, 1H, $J = 2.2$ Hz); 7.30 (*dd*, 1H, $J_1 = 8.90$ Hz, $J_2 = 2.18$ Hz); 7.24 (*m*, 1H); 6.90 (*d*, 1H, $J = 7.8$ Hz); 6.81 (*m*, 2H); 4.78 (*dd*, 1H, $J_1 = 12.9$ Hz, $J_2 = 9.7$ Hz); 4.69 (*dd*, 1H, $J_1 = 12.9$ Hz, $J_2 = 5.7$ Hz); 4.06 (*s*, 3H); 3.99 (*dd*, 1H, $J_1 = 12.2$ Hz, $J_2 = 7.9$ Hz); 3.81 (*m*, 4H); 2.85 (*dd*, 1H, $J_1 = 12.1$ Hz, $J_2 = 7.4$ Hz); 2.65 (*m*, 1H); 2.25 (*m*, 1H); 1.55 (*m*, 1H); 1.05 (*t*, 3H, $J = 7.4$ Hz). ^{13}C-NMR (150 MHz, $CDCl_3$): δ (ppm): 165.1, 163.3, 159.9, 151.4, 140.8, 136.2, 130.0, 129.9, 128.5, 124.1, 123.4, 122.6, 119.6, 116.0, 113.4, 113.1, 79.4, 55.2, 54.1, 43.7, 38.5, 38.4, 20.3, 14.3.

(*R*,*Z*)-methyl-3-(2-(2-(benzyloxy)phenyl)-1-nitroheptan-4-ylidene)-5-chloro-2-oxoindoline-1-carboxylate (product **110**, Table 2.3—entry 5)

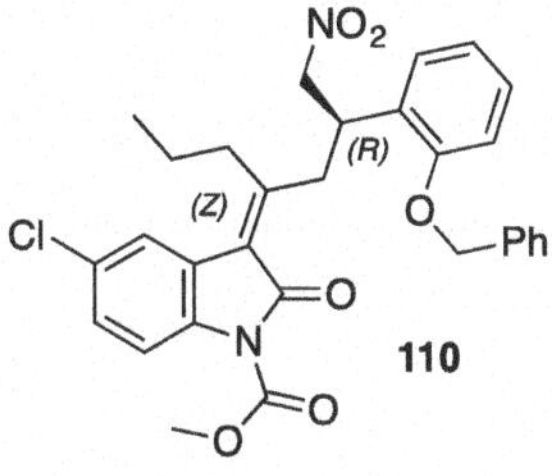

110

The reaction was carried out following the general procedure to furnish the crude product **110** as 93:7 mixture of (Z)-**110** and (*E*)-**110**. The crude mixture has been purified by flash column chromatography (hexane/ethyl acetate = 8/2) to give an overall yield of 93% and (Z)-**110** in a >99% ee. HPLC analysis on a AD-H column: hexane/*i*-PrOH 95/5, flow rate 0.5 mL/min, $\lambda = 254$ nm: $\tau_{major} = 38.12$ min. $[\alpha]_D^{20}$ −72.8 (*c* 1.00, $CHCl_3$). HRMS-ESI (+): calculated for $C_{30}H_{29}ClNaN_2O_6$ 571.1606, found 571.1602 $[M+Na]^+$. ^1H NMR (600 MHz, $CDCl_3$): δ (ppm): 7.91 (*d*, 1H, *J* = 9.2 Hz); 7.37 (*m*, 6H); 7.24 (*m*, 3H); 6.93 (*m*, 2H); 5.08 (*d*, 1H, *J* = 11.2 Hz); 4.98 (*m*, 2H); 4.70 (*dd*, 1H, $J_1 = 12.9$ Hz, $J_2 = 6.2$ Hz); 4.36 (*m*, 1H); 4.03 (*s*, 3H); 3.83 (*dd*, 1H, $J_1 = 11.2$ Hz, $J_2 = 8.4$ Hz); 2.93 (*m*, 1H); 2.50 (*m*, 1H); 2.14 (*m*, 1H); 1.44 (*m*, 2H); 0.90 (*t*, 3H, *J* = 7.2 Hz). ^{13}C-NMR (150 MHz, $CDCl_3$): δ (ppm): 164.9, 164.3, 156.2, 151.4, 136.6, 136.1, 129.7, 128.8, 128.6, 128.2, 128.0, 127.4, 124.3, 123.3, 122.2, 121.3, 116.0, 112.2, 78.0, 70.4, 53.9, 38.3, 37.4, 29.7, 20.2, 14.2.

(*R*,*Z*)-methyl-5-chloro-3-(1-nitro-2-(p-tolyl)heptan-4-ylidene)-2-oxoindoline-1-carboxylate (product **111**, Table 2.3—entry 6)

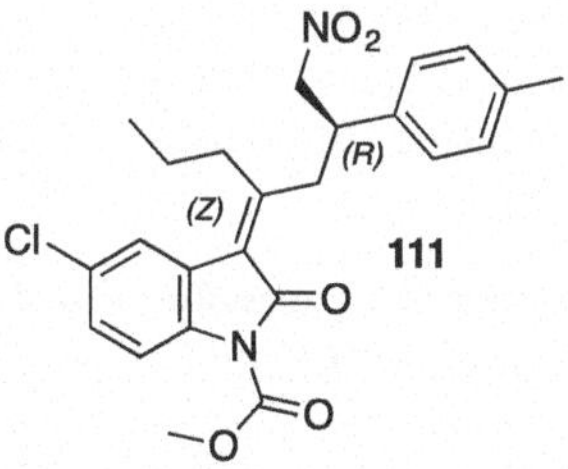

111

The reaction was carried out following the general procedure to furnish the crude product **111** as 94:6 mixture of (Z)-**111** and (*E*)-**111**. The crude mixture has been purified by flash column chromatography (hexane/ethyl acetate = 8/2) to give an overall yield of 96% and (Z)-**11** in a >99% ee. HPLC analysis on a AD-H column: hexane/*i*-PrOH 95/5, flow rate 1 mL/min, $\lambda = 254$ nm: $\tau_{major} = 10.61$ min. $[\alpha]_D^{20}$ −57.0 (*c* 1.00, $CHCl_3$). HRMS-ESI (+): calculated for $C_{24}H_{25}ClNaN_2O_5$ 479.1344, found 479.1339 $[M+Na]^+$. ^1H NMR (600 MHz, $CDCl_3$): δ (ppm): 7.94 (*d*, 1H, *J* = 8.9 Hz); 7.40 (*d*, 1H, *J* = 1.9 Hz); 7.29 (*dd*, 1H, $J_1 = 8.9$ Hz, $J_2 = 1.9$ Hz); 7.19

(*d*, 1H, J_1 = 8.1 Hz); 7.13 (*d*, 1H, J_1 = 8.1 Hz); 4.76 (*dd*, 1H, J_1 = 12.7 Hz, J_2 = 10.0 Hz); 4.68 (*dd*, 1H, J_1 = 12.7 Hz, J_2 = 5.6 Hz); 4.06 (*s*, 3H); 4.01 (*m*, 1H); 3.81 (*m*, 1H); 2.81 (*dd*, 1H, J_1 = 12.1 Hz, J_2 = 7.5 Hz); 2.64 (*m*, 1H); 2.28 (*m*, 4H); 1.55 (*m*, 2H); 1.05 (*t*, 3H, J_1 = 7.2 Hz). ^{13}C-NMR (150 MHz, $CDCl_3$): δ (ppm): 165.0, 163.6, 151.3, 137.6, 136.2, 136.1, 129.8, 129.6, 128.4, 127.3, 124.1, 123.4, 122.5, 116.0, 79.5, 54.0, 43.3, 38.7, 38.3, 31.5, 22.6, 21.0, 20.2, 14.3, 14.1.

(*R*,*Z*)-methyl-5-chloro-3-(2-(2-fluorophenyl)-1-nitroheptan-4-ylidene)-2-oxoindoline-1-carboxylate (product **112**, Table 2.3—entry 7)

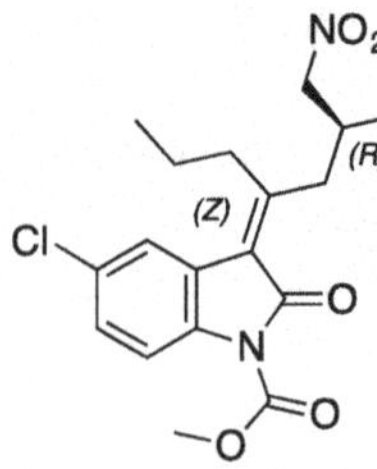

112

The reaction was carried out following the general procedure to furnish the crude product **112** as 95:5 mixture of (*Z*)-**112** and (*E*)-**112**. The crude mixture has been purified by flash column chromatography (hexane/ethyl acetate = 8/2) to give an overall yield of 87% and (*Z*)-**112** in a >99% ee. HPLC analysis on a AD-H column: hexane/*i*-PrOH 95/5, flow rate 1.0 mL/min, λ = 254 nm: τ_{major} = 12.92 min. $[\alpha]_D^{20}$ −96.2 (*c* 1.00, $CHCl_3$). HRMS-ESI (+): calculated for $C_{23}H_{22}ClFNaN_2O_5$ 483.1093, found 483.1089 $[M+Na]^+$. ^{1}H NMR (600 MHz, $CDCl_3$): *δ* (ppm): 7.95 (*d*, 1H, *J* = 8.9 Hz); 7.40 (*bs*, 1H); 7.28 (*m*, 3H); 7.08 (*m*, 2H); 4.91 (*m*, 1H); 4.76 (*dd*, 1H, J_1 = 12.5 Hz, J_2 = 5.5 Hz); 4.15 (*m*, 1H); 4.06 (*s*, 3H); 3.94 (*m*, 1H); 2.92 (*m*, 1H); 2.65 (*m*, 1H); 2.26 (*m*, 1H); 1.55 (*m*, 2H); 1.05 (*t*, 3H, *J* = 7.1 Hz). ^{13}C NMR (150 MHz, $CDCl_3$): δ (ppm): 165.0, 162.9, 160.7 (d, *J* = 246.5 Hz), 151.4, 136.3, 129.9, 129.6 (d, *J* = 8.5 Hz), 129.3 (d, *J* = 4.4 Hz), 128.5, 126.0 (d, *J* = 13.6 Hz), 124.7 (d, *J* = 3.7 Hz), 124.1, 123.4, 122.7, 116.1, 116.0 (d, *J* = 22.1 Hz), 77.9, 54.0, 38.4, 37.9, 37.4, 29.7, 20.2, 14.3.

(*R*,*Z*)-methyl-3-(2-(4-bromophenyl)-1-nitroheptan-4-ylidene)-5-chloro-2-oxoindoline-1-carboxylate (product **113**, Table 2.3—entry 8)

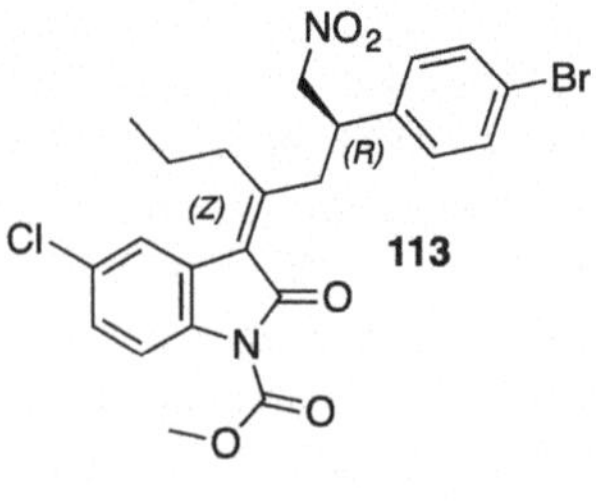

113

The title compound was obtained as single diastereosiomer. After purification by flash column chromatography (hexane/ethyl acetate = 8/2) (Z)-**113** was obtained in 85% yield and >99% ee. HPLC analysis on a AD-H column: hexane/*i*-PrOH 95/5, flow rate 1 mL/min, λ = 214 nm: τ_{major} = 14.15 min. $[\alpha]_D^{20}$ −60.0 (*c* 1.00, $CHCl_3$). HRMS-ESI (+): calculated for $C_{23}H_{22}BrClNaN_2O_5$ 543.0298, found 543.0295 $[M+Na]^+$. ^{1}H NMR (400 MHz, $CDCl_3$): δ (ppm): 7.95 (*d*, 1H, J = 8.8 Hz); 7.46 (*d*, 2H, J = 8.3 Hz); 7.41 (*d*, 1H, J = 2.0 Hz); 7.31 (*dd*, 1H, J_1 = 8.7 Hz, J_2 = 2.0 Hz); 7.20 (*d*, 2H, J = 8.4 Hz); 4.75 (*dd*, 1H, J_1 = 13.0 Hz, J_2 = 10.0 Hz); 4.68 (*dd*, 1H, J_1 = 13.0 Hz, J_2 = 5.5 Hz); 4.07 (*s*, 3H); 3.95 (*dd*, 1H, J_1 = 12.2 Hz, J_2 = 8.2 Hz); 3.83 (*m*, 1H); 2.84 (*dd*, 1H, J_1 = 12.2 Hz, J_2 = 7.4 Hz); 2.65 (*m*, 2H); 2.26 (*m*, 2H); 1.57 (*m*, 2H); 1.07 (*t*, 3H, J = 7.3 Hz). ^{13}C NMR (100 MHz, $CDCl_3$): δ (ppm): 164.1, 161.5, 150.3, 137.3, 135.3, 131.2, 129.0, 128.2, 127.7, 123.0,122.5, 121.9, 120.9, 115.1, 78.05, 53.1, 42.1, 37.5, 37.4, 19.3, 13.3.

(*R*,*Z*)-methyl-5-chloro-3-(2-(2,6-dichlorophenyl)-1-nitroheptan-4-ylidene)-2-oxoindoline-1-carboxylate (product **114**, Table 2.3—entry 9)

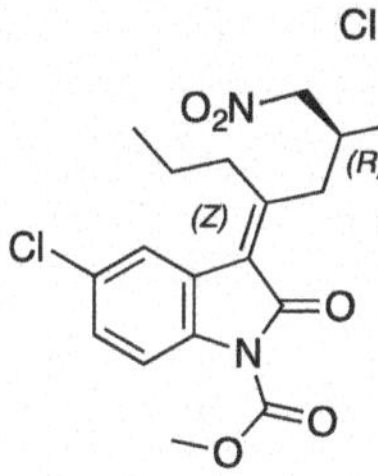

114

The reaction was carried out following the general procedure to furnish the crude product **114** as 92:8 mixture of (*Z*)-**114** and (*E*)-**114**. The crude mixture has been purified by flash column chromatography (hexane/ethyl acetate = 8/2) to give an overall yield of 80% and (Z)-**114** in a >99% ee. HPLC analysis on a AD-H column: hexane/*i*-PrOH 95/5, flow rate 0.3 mL/min, λ = 254 nm: **114** τ_{major} = 55.94 min. $[\alpha]_D^{20}$ −125.9 (*c* 1.00, $CHCl_3$). HRMS-ESI (+): calculated for $C_{23}H_{21}Cl_3NaN_2O_5$ 533.0408, found 533.0410 $[M+Na]^+$. ^{1}H NMR (600 MHz, $CDCl_3$): δ (ppm): 7.96 (*d*, 1H, J = 9.6 Hz); 7.40 (*m*, 1H); 7.31 (*m*, 3H); 7.16 (*m*, 1H); 5.40 (*m*, 1H); 4.94 (*m*, 2H); 4.27 (*m*, 1H); 4.05 (*s*, 3H); 2.90 (*m*, 1H); 2.69 (*m*, 1H); 2.15 (*m*, 1H); 1.54 (*m*, 2H); 1.03 (*t*, 3H, J = 7.1 Hz). ^{13}C-NMR (150 MHz, $CDCl_3$): δ (ppm): 164.9, 162.4, 151.4, 136.9, 136.4, 134.8, 134.3, 130.2, 129.8, 129.4, 129.1, 128.5, 124.1, 123.4, 122.8, 116.0, 76.5, 54.1, 39.2, 38.6, 35.0, 29.7, 20.2, 14.3.

(*S*,*Z*)-methyl-5-chloro-3-(8-methyl-6-(nitromethyl)nonan-4-ylidene)-2-oxoindoline-1-carboxylate (product **115**, Table 2.3—entry 10)

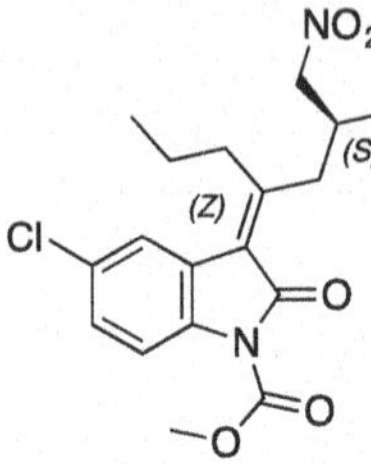

115

The title compound was obtained as single diastereosiomer. After purification by flash column chromatography (hexane/ethyl acetate = 85/15) (Z)-**115** was obtained in 50% yield and 97% ee. HPLC analysis on a amylose-2 column: hexane/*i*-PrOH 95/5, flow rate 1.0 mL/min, $\lambda = 254$ nm: **115** $\tau_{major} = 18.74$ min; $\tau_{minor} = 14.2$ min. $[\alpha]_D^{20}$ +155.0 (*c* 1.00, $CHCl_3$). HRMS-ESI (+): calculated for $C_{21}H_{27}ClNaN_2O_5$ 445.1501, found 445.1501 $[M+Na]^+$. ^1H NMR (600 MHz, $CDCl_3$): δ (ppm): 7.95 (*d*, 1H, $J = 8.9$ Hz); 7.45 (*d*, 1H, $J = 1.7$ Hz); 7.31 (*dd*, 1H, $J_1 = 8.9$ Hz, $J_2 = 1.7$ Hz); 4.42 (*dd*, 1H, $J_1 = 12.7$ Hz, $J_2 = 7.2$ Hz); 4.33 (*dd*, 1H, $J_1 = 12.7$ Hz, $J_2 = 6.1$ Hz); 4.04 (*s*, 3H); 3.48 (*dd*, 1H, $J_1 = 12.2$ Hz, $J_2 = 9.2$ Hz); 2.84 (*dd*, 1H, $J_1 = 12.2$ Hz, $J_2 = 6.1$ Hz); 2.72 (*m*, 2H); 2.57 (*m*, 1H); 1.69 (*m*, 4H); 1.35 (*m*, 1H); 1.14 (*t*, 3H, $J = 7.2$ Hz); 0.93(*m*, 6H). ^{13}C NMR (150 MHz, $CDCl_3$): δ (ppm): 165.2, 164.1, 151.4, 136.1, 129.9, 128.4, 123.4, 122.9, 120.3, 116.0, 79.3, 54.0, 41.4, 38.2, 36.8, 35.0, 25.1, 22.6, 22.3, 20.6, 14.4.

(*R*,*Z*)-methyl-6-chloro-3-(1-nitro-2-phenylheptan-4-ylidene)-2-oxoindoline-1-carboxylate (product **116**, Table 2.3—entry 11)

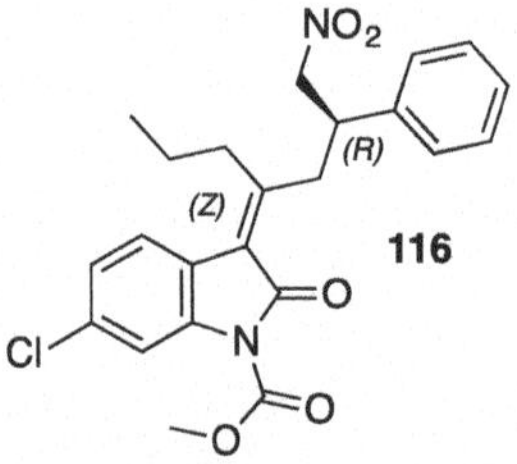

116

The reaction was carried out following the general procedure to furnish the crude product **116** as 92:8 mixture of (Z)-**116** and (*E*)-**116**. The crude mixture has been purified by flash column chromatography (hexane/ethyl acetate = 8/2) to give an overall yield of 75% and (Z)-**116** in a >99% ee. HPLC analysis on a AD-H column: hexane/*i*-PrOH 95/5, flow rate 1.0 mL/min, $\lambda = 214$ nm: $\tau_{major} = 24.60$ min; $\tau_{minor} = 16.69$ min. $[\alpha]_D^{20}$ −102.0 (*c* 1.00, $CHCl_3$). HRMS-ESI (+): calculated for $C_{23}H_{23}ClNaN_2O_5$ 465.1188, found 465.1188 $[M+Na]^+$. ^1H NMR (600 MHz, $CDCl_3$): δ (ppm): 8.04 (*d*, 1H, $J = 1.7$ Hz); 7.32 (*m*, 6H); 7.17 (*dd*, 1H, $J_1 = 8.4$ Hz,

$J_2 = 1.7$ Hz); 4.80 (*dd*, 1H, $J_1 = 12.6$ Hz, $J_2 = 9.8$ Hz); 4.70 (*dd*, 1H, $J_1 = 12.9$ Hz, $J_2 = 5.5$ Hz); 4.07 (*s*, 3H); 4.03 (*m*, 1H); 3.85 (*m*, 1H); 2.80 (*dd*, 1H, $J_1 = 12.2$ Hz, $J_2 = 7.3$ Hz); 2.64 (*m*, 1H); 2.24 (*m*, 1H); 1.53 (*m*, 2H); 1.03 (*t*, 3H, $J = 7.3$ Hz). ^{13}C-NMR (150 MHz, $CDCl_3$): δ (ppm): 165.3, 161.9, 151.3, 139.3, 138.6, 134.6, 128.9, 127.9, 127.5, 124.5, 124.0, 122.7, 121.2, 115.5, 79.4, 54.1, 43.6, 38.6, 38.5, 20.2, 14.4.

(*Z*)-methyl 3-((2*R*,3*R*)-3-methyl-1-nitro-2-phenylheptan-4-ylidene)-2-oxoindoline-1-carboxylate (product **117**, Table 2.3—entry 12)

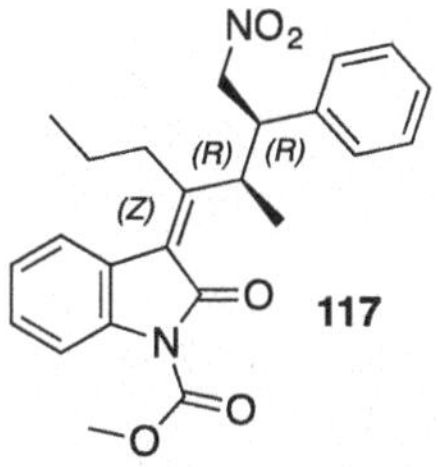

117

The title compound was obtained as single diastereoisomer. After purification by flash column chromatography (hexane/ethyl acetate = 8/2) (Z)-**117** was obtained in 44% yield and >99% ee. HPLC analysis on a AD-H column: hexane/*i*-PrOH 95/5, flow rate 0.5 mL/min, $\lambda = 254$ nm: $\tau_{major} = 22.36$ min. $[\alpha]_D^{20}$ +145.5 (*c* 1.00, $CHCl_3$). HRMS-ESI (+): calculated for $C_{24}H_{26}NaN_2O_5$ 445.1734, found 445.1732 $[M+Na]^+$. ^{1}H NMR (400 MHz, $CDCl_3$): δ (ppm): 7.91 (*d*, 1H, $J = 7.9$ Hz); 7.25 (*m*, 1H); 7.15 (*m*, 5H); 7.09 (*m*, 2H); 5.45 (*m*, 1H); 4.85 (*dd*, 1H, $J_1 = 12.2$ Hz, $J_2 = 4.3$ Hz); 4.64 (*dd*, 1H, $J_1 = 12.1$ Hz, $J_2 = 10.2$ Hz); 4.06 (*s*, 3H); 3.70 (*m*, 1H); 2.43 (*m*, 1H); 2.32 (*m*, 1H); 1.47 (*m*, 2H); 1.34 (*d*, 1H, $J = 6.5$ Hz); 1.06 (*t*, 3H, $J = 7.2$ Hz). ^{13}C NMR (100 MHz, $CDCl_3$): δ (ppm): 166.1, 165.8, 151.6, 138.2, 137.4, 128.6, 128.4, 127.8, 127.7, 124.2, 123.3, 122.9, 122.8, 114.5, 80.0, 53.9, 48.8, 36.4, 33.3, 21.4, 17.1, 14.6.

(*R*,*Z*)-methyl-3-(7-methyl-1-nitro-2-phenyloct-7-en-4-ylidene)-2-oxoindoline-1-carboxylate (product **118**, Table 2.3—entry 13)

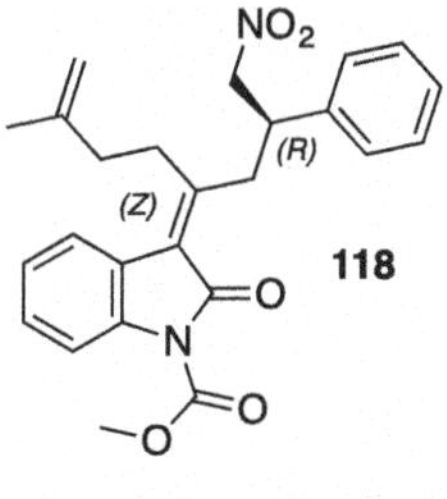

118

The reaction was carried out following the general procedure to furnish the crude product **118** as 80:20 mixture of (Z)-**118** and (*E*)-**118**. The crude mixture has been purified by flash column chromatography (hexane/ethyl acetate = 8/2) to give an overall yield of 92% and (Z)-**118** in a >99% ee. HPLC analysis on a cellulose-2 column: hexane/*i*-PrOH 90/10, flow rate 0.5 mL/min, $\lambda = 214$ nm: $\tau_{major} = 39.79$ min. $[\alpha]_D^{20}$ −100.8 (*c* 1.00, $CHCl_3$). HRMS-ESI (+): calculated for $C_{25}H_{26}NaN_2O_5$ 457.1734, found 457.1730 $[M+Na]^+$. ^1H NMR (600 MHz, $CDCl_3$): δ (ppm): 8.00 (*d*, 1H, *J* = 8.2 Hz); 7.49 (*m*, 1H); 7.27 (*m*, 6H); 4.77 (*m*, 4H); 4.07 (*s*, 3H); 3.87 (*m*, 1H); 2.76 (*m*, 1H); 2.41 (*m*, 1H); 2.14(*m*, 1H); 1.74 (*s*, 3H). ^{13}C NMR (150 MHz, $CDCl_3$): δ (ppm): 164.5, 158.3, 150.6, 141.6, 136.8, 136.4, 127.5, 127.4, 126.9, 123.2, 123.0, 122.9, 122.7, 113.5, 112.2, 78.6, 52.9, 47.5, 39.8, 39.2, 21.2, 17.8.

(*R*,*Z*)-methyl 3-(6-nitro-1,5-diphenylhexan-3-ylidene)-2-oxoindoline-1-carboxylate (product **119**, Table 2.3—entry 14)

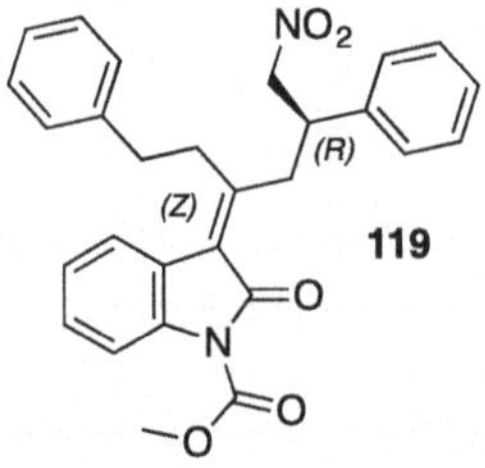

119

The reaction was carried out following the general procedure to furnish the crude product **119** as 80:20 mixture of (Z)-**119** and (*E*)-**119**. The crude mixture has been purified by flash column chromatography (hexane/ethyl acetate = 8/2) to give an overall yield of 63% and (Z)-**119** in a >99% ee. HPLC analysis on a cellulose-2 column: hexane/*i*-PrOH 90/10, flow rate 0.5 mL/min, $\lambda = 214$ nm: $\tau_{major} = 49.28$ min. $[\alpha]_D^{20}$ −92.5 (*c* 1.00, $CHCl_3$). HRMS-ESI (+): calculated for $C_{28}H_{26}NaN_2O_5$ 493.1734, found 493.1735 $[M+Na]^+$. ^1H NMR (600 MHz, $CDCl_3$): δ (ppm): 8.02 (*d*, 1H, *J* = 8.7 Hz); 7.58 (*d*, 1H, *J* = 7.7 Hz); 7.26 (*m*, 12H); 4.79 (*dd*, 1H, J_1 = 12.8 Hz, J_2 = 9.8 Hz); 4.73 (*dd*, 1H, J_1 = 12.8 Hz, J_2 = 5.5 Hz); 4.07 (*s*, 3H); 3.99 (*dd*, 1H, J_1 = 12.2 Hz, J_2 = 7.6 Hz); 3.85 (*m*, 1H); 3.02 (*m*, 1H); 2.75 (*m*, 3H); 2.56 (*m*, 3H). ^{13}C NMR (150 MHz, $CDCl_3$): δ (ppm): 164.6, 158.9, 150.5, 138.9, 138.4, 137.0, 128.2, 128.1, 128.0, 127.8, 127.0, 126.9, 126.5, 126.2, 125.6, 123.5, 122.8, 121.6, 114.1, 78.45, 53.0, 42.8, 37.7, 36.9, 31.4.

(*Z*)-methyl-5-chloro-3-((3*R*,4*R*)-3-ethyl-5-nitro-4-phenylpentan-2-ylidene)-2-oxoindoline-1-carboxylate (product **120**, Table 2.4—entry 1)

120

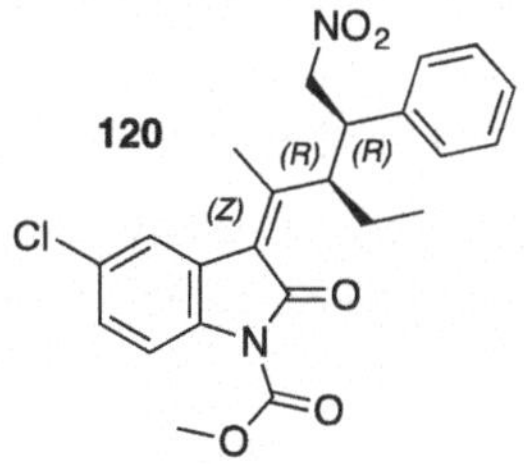

The title compound was obtained as single diastereoisomer. After purification by flash column chromatography (hexane/ethyl acetate = 8/2) (Z)-**120** was obtained in 86% yield and >99% ee. HPLC analysis on a AD-H column: hexane/*i*-PrOH 90/10, flow rate 0.5 mL/min, λ = 214 nm: τ_{major} = 14.10. $[\alpha]_D^{20}$ +117.1 (*c* 1.00, $CHCl_3$). HRMS-ESI (+): calculated for $C_{23}H_{23}ClNaN_2O_5$ 465.1188, found 465.1188 $[M+Na]^+$. ^1H NMR (600 MHz, $CDCl_3$): δ (ppm): 7.87 (*d*, 1H, *J* = 8.7 Hz); 7.32 (*d*, 1H, *J* = 2.0 Hz); 7.23 (*dd*, 1H, J_1 = 8.8 Hz, J_2 = 2.0 Hz); 7.17 (*m*, 4H); 7.11 (*m*, 1H); 5.28 (*ddd*, 1H, J_1 = 10.7 Hz, J_2 = 4.2 Hz); 4.88 (*dd*, 1H, J_1 = 12.4 Hz, J_2 = 4.4 Hz); 4.66 (*dd*, 1H, J_1 = 12.5 Hz, J_2 = 10.5 Hz); 4.05 (*s*, 3H); 3.65 (*ddd*, 1H, J_1 = 10.6 Hz, J_2 = 4.7 Hz); 2.03 (*s*, 3H); 1.86 (*m*, 1H); 1.68 (*m*, 1H); 0.85 (*t*, 3H, *J* = 7.4 Hz). ^{13}C NMR (150 MHz, $CDCl_3$): δ (ppm): 164.0 (C); 161.0 (C); 150.5 (C); 137.0 (C); 134.7 (C); 128.6 (C); 127.6 (CH); 127.1 (CH); 126.9 (CH); 126.7 (CH); 123.9 (C); 122.8 (CH); 114.7 (CH); 79.1 (CH_2); 53.0 (CH_3); 47.01 (CH); 42.4 (CH); 23.5 (CH_2); 17.25 (CH_3); 10.7 (CH_3).

(*Z*)-methyl-5-chloro-3-((3*R*,4*S*)-3-ethyl-5-nitro-4-(thiophen-2-yl)pentan-2-ylidene)-2-oxoindoline-1-carboxylate (product **121**, Table 2.4—entry 2)

121

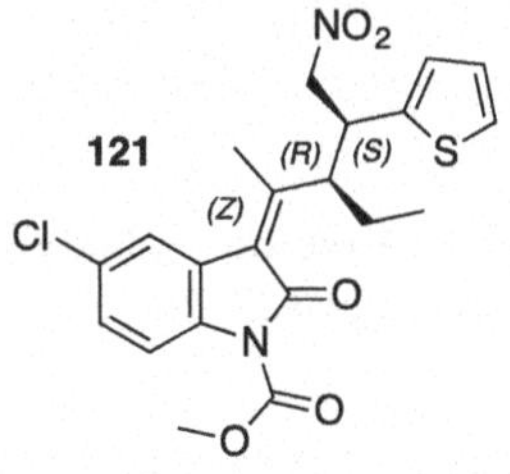

The title compound was obtained as single diastereoisomer. After purification by flash column chromatography (hexane/ethyl acetate = 8/2) (Z)-**121** was obtained in 89% yield and 99% ee. HPLC analysis on a AD-H column: hexane/*i*-PrOH 90/10, flow rate 0.5 mL/min, λ = 254 nm: τ_{major} = 17.31 min; τ_{minor} = 25.85 min. $[\alpha]_D^{20}$ +72.9 (*c* 1.00, $CHCl_3$). HRMS-ESI (+): calculated for $C_{21}H_{21}ClNaN_2O_5S$ 471.0752, found 471.0750 $[M+Na]^+$. ^1H NMR (400 MHz, $CDCl_3$): δ (ppm): 7.91 (*d*, 1H, *J* = 8.7 Hz); 7.42 (*d*, 1H, *J* = 2.0 Hz); 7.26 (*dd*, 1H, J_1 = 8.7 Hz, J_2 = 2.0 Hz);

7.07 (*d*, 1H, J = 5.2 Hz); 6.87 (*d*, 1H, J = 3.4 Hz); 6.80 (*dd*, 1H, J_1 = 5.0 Hz, J_2 = 3.6 Hz); 5.26 (*ddd*, 1H, J_1 = 10.4 Hz, J_2 = 4.3 Hz); 4.90 (*dd*, 1H, J_1 = 12.6 Hz, J_2 = 4.6 Hz); 4.65 (*dd*, 1H, J_1 = 12.6 Hz, J_2 = 10.4 Hz); 4.03 (*m*, 4H); 2.12 (*s*, 3H); 1.85 (*m*, 1H); 1.69 (*m*, 1H); 0.86 (*t*, 3H, J = 7.3 Hz). ^{13}C NMR (100 MHz, $CDCl_3$): δ (ppm): 164.9, 161.17, 151.5, 140.2, 135.8, 129.6, 128.2, 126.8, 126.0, 125.0, 124.9, 124.2, 123.9, 115.7, 80.2, 54.0, 44.3, 42.8, 30.9, 24.4, 18.2, 11.6.

(Z)-methyl-3-((3*R*,4*R*)-3-ethyl-5-nitro-4-phenylpentan-2-ylidene)-2-oxoindoline-1-carboxylate (product **122**, Table 2.4—entry 3)

122

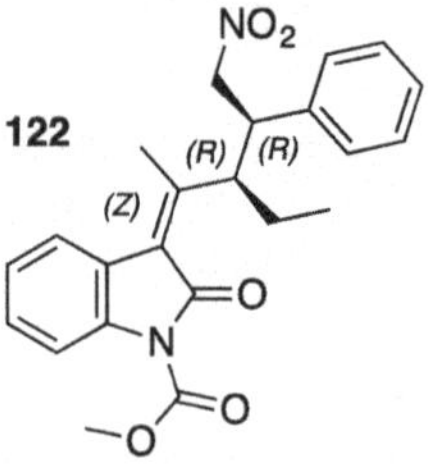

The title compound was obtained as single diastereoisomer. After purification by flash column chromatography (hexane/ethyl acetate = 8/2) (Z)-**122** was obtained in 70% yield and > 99% ee. HPLC analysis on a AD-H column: hexane/*i*-PrOH 90/10, flow rate 0.5 mL/min, λ = 214 nm: τ_{major} = 17.7. $[\alpha]_D^{20}$ +147.1 (*c* 1.00, $CHCl_3$). HRMS-ESI (+): calculated for $C_{23}H_{24}NaN_2O_5$ 431.1577, found 431.1574 $[M+Na]^+$. ^{1}H NMR (600 MHz, $CDCl_3$): δ (ppm): 7.92 (*d*, 1H, J = 8.2 Hz); 7.36 (*d*, 1H, J_1 = 7.8 Hz); 7.27 (*m*, 1H); 7.20 (*m*, 2H); 7.15 (*m*, 2H); 7.09 (*m*, 2H); 5.30 (*ddd*, 1H, J_1 = 10.7 Hz, J_2 = 4.1 Hz); 4.89 (*dd*, 1H, J_1 = 12.5 Hz, J_2 = 4.4 Hz); 4.61 (*dd*, 1H, J_1 = 12.4 Hz, J_2 = 10.4 Hz); 4.05 (*s*, 3H); 3.65 (*ddd*, 1H, J_1 = 10.4 Hz, J_2 = 4.5 Hz); 2.03 (*s*, 3H); 1.86 (*m*, 1H); 1.68 (*m*, 1H); 0.86 (*t*, 3H, J = 7.3 Hz). ^{13}C NMR (150 MHz, $CDCl_3$): δ (ppm): 164.6 (C); 158.7 (C); 150.7 (C); 137.2 (C); 136.3 (C); 127.5 (CH); 127.4 (CH); 126.8 (CH); 123.6 (CH); 123.0 (CH); 122.8 (C); 122.6 (C); 113.6 (CH); 79.2 (CH_2); 52.9 (CH_3); 47.1 (CH); 42.2 (CH); 23.5 (CH_2); 17.1 (CH_3); 10.7 (CH_3).

(Z)-methyl-3-((3*R*,4*R*)-3-ethyl-4-(4-methoxyphenyl)-5-nitropentan-2-ylidene)-2-oxoindoline-1-carboxylate (product **123**, Table 2.4—entry 4)

123

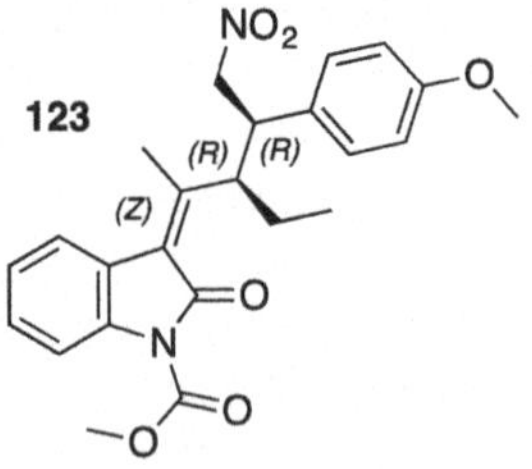

The title compound was obtained as single diastereoisomer. After purification by flash column chromatography (hexane/ethyl acetate = 8/2) (*Z*)-**123** was obtained in 52% yield and 98% ee. HPLC analysis on a AD-H column: hexane/*i*-PrOH 90/10, flow rate 0.5 mL/min, $\lambda = 214$ nm: $\tau_{major} = 19.9$ min. $[\alpha]_D^{20}$ +200.0 (*c* 1.00, $CHCl_3$). HRMS-ESI (+): calculated for $C_{24}H_{26}NaN_2O_6$ 461.1683, found 461.1677 $[M+Na]^+$. ^{1}H NMR (600 MHz, $CDCl_3$): δ (ppm): 7.93 (*d*, 1H, J = 8.2 Hz); 7.39 (*d*, 1H, J = 7.3 Hz); 7.28 (*m*, 1H); 7.11 (*m*, 3H); 6.68 (*d*, 2H, J = 8.5 Hz); 5.27 (*ddd*, 1H, J_1 = 10.6 Hz, J_2 = 4.3 Hz); 4.86 (*dd*, 1H, J_1 = 12.2 Hz, J_2 = 4.3 Hz); 4.61 (*dd*, 1H, J_1 = 12.1 Hz, J_2 = 10.8 Hz); 4.06 (*s*, 3H); 3.67 (*s*, 3H); 3.60 (*ddd*, 1H, J_1 = 10.9 Hz, J_2 = 4.6 Hz); 2.03 (*s*, 3H); 1.83 (*m*, 1H); 1.66 (*m*, 1H); 0.85 (*t*, 3H, J = 7.4 Hz). ^{13}C NMR (150 MHz, $CDCl_3$): δ (ppm): 164.6 (C); 159.1 (C); 157.8 (C); 150.7 (C); 136.3 (C); 129.1 (C); 127.8 (CH); 127.4 (CH); 123.5 (C); 123.0 (CH); 122.8 (CH); 122.7 (C); 113.6 (CH); 112.9 (CH); 79.4 (CH_2); 54.0 (CH_3); 52.9 (CH_3); 46.4 (CH); 42.2 (CH); 23.6 (CH_2); 17.1 (CH_3); 10.7 (CH_3).

(*Z*)-methyl-3-((3*R*,4*R*)-3-ethyl-4-(3-methoxyphenyl)-5-nitropentan-2-ylidene)-2-oxoindoline-1-carboxylate (product **124**, Table 2.4—entry 5)

124

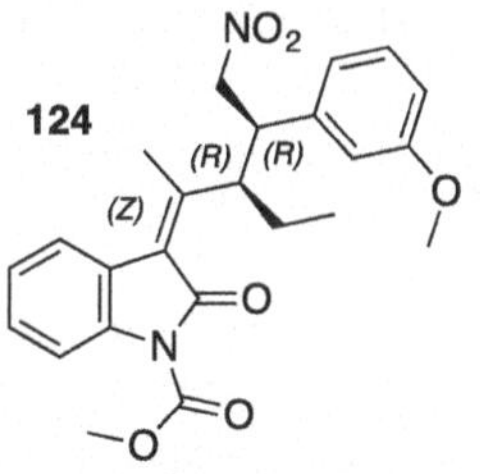

The title compound was obtained as single diastereoisomer. After purification by flash column chromatography (hexane/ethyl acetate = 8/2) (*Z*)-**124** was obtained in 61% yield and >99% ee. HPLC analysis on a AD-H column: hexane/*i*-PrOH 90/10, flow rate 0.5 mL/min, $\lambda = 214$ nm: $\tau_{major} = 21.39$ min. $[\alpha]_D^{20}$ +105.4 (*c* 1.00, $CHCl_3$). HRMS-ESI (+): calculated for $C_{24}H_{26}N_2O_6$ 461.1683, found 461.1677 $[M+Na]^+$. ^{1}H NMR (600 MHz, $CDCl_3$): δ (ppm): 7.92 (*d*, 1H, J = 8.1 Hz); 7.39 (*d*, 1H, J = 7.9 Hz); 7.27 (*m*, 1H); 7.10 (*m*, 1H); 7.05 (*m*, 1H); 6.76 (*m*, 2H); 6.63 (*dd*, 1H, J_1 = 8.3 Hz, J_2 = 2.6 Hz); 5.30 (*ddd*, 1H, J_1 = 10.5 Hz, J_2 = 4.1 Hz); 4.87 (*dd*, 1H, J_1 = 12.5 Hz, J_2 = 4.5 Hz); 4.65 (*dd*, 1H, J_1 = 12.4 Hz, J_2 = 10.3 Hz); 4.05 (*s*, 3H); 3.68 (*s*, 3H); 3.64 (*ddd*, 1H, J_1 = 10.7 Hz, J_2 = 4.8 Hz); 2.06 (*s*, 3H); 1.84 (*m*, 1H); 1.67 (*m*, 1H); 0.85 (*t*, 3H, J = 7.5 Hz). ^{13}C NMR (100 MHz, $CDCl_3$): δ (ppm): 164.6 (C); 158.8 (C); 158.5 (C); 150.6 (C); 138.7 (C); 136.3 (C); 128.6 (CH); 127.4 (CH); 123.6 (C); 123.1 (CH); 122.8 (CH); 122.7 (C); 119.1 (CH); 113.6 (CH); 112.7 (CH); 112.0 (CH); 79.2 (CH_2); 54.0 (CH_3); 52.8 (CH_3); 47.1 (CH); 42.1 (CH); 23.5 (CH_2); 17.1 (CH_3); 10.7 (CH_3).

(Z)-methyl-3-((3*R*,4*R*)-3-ethyl-5-nitro-4-(p-tolyl)pentan-2-ylidene)-2-oxoindoline-1-carboxylate (product **125**, Table 2.4—entry 6)

125

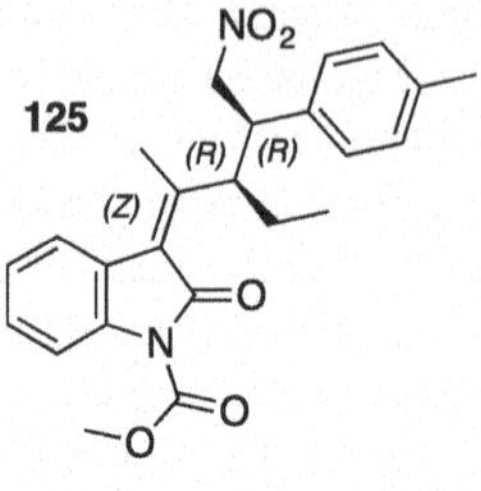

The title compound was obtained as single diastereoisomer. After purification by flash column chromatography (hexane/ethyl acetate = 8/2) (Z)-**125** was obtained in 75% yield and >99% ee. HPLC analysis on a AD-H column: hexane/*i*-PrOH 90/10, flow rate 0.5 mL/min, $\lambda = 214$ nm: $\tau_{major} = 15.93$ min. $[\alpha]_D^{20}$ +155.0 (*c* 1.00, $CHCl_3$). HRMS-ESI (+): calculated for $C_{24}H_{26}NaN_2O_5$ 445.1734, found 445.1733 $[M+Na]^+$. ^1H NMR (600 MHz, $CDCl_3$): δ (ppm): 7.94 (*d*, 1H, $J = 8.2$ Hz); 7.39 (*d*, 1H, $J = 7.8$ Hz); 7.27 (*ddd*, 1H, $J = 7.5$ Hz); 7.09 (*ddd*, 1H, $J = 8.0$ Hz); 7.07 (*d*, 2H, $J = 8.0$ Hz); 6.69 (*d*, 2H, $J = 7.8$ Hz); 5.28 (*ddd*, 1H, $J_1 = 10.5$ Hz, $J_2 = 4.0$ Hz); 4.86 (*dd*, 1H, $J_1 = 12.2$ Hz, $J_2 = 4.4$ Hz); 4.62 (*dd*, 1H, $J_1 = 12.2$ Hz, $J_2 = 10.6$ Hz); 4.06 (*s*, 3H); 3.62 (*ddd*, 1H, $J_1 = 10.4$ Hz, $J_2 = 4.4$ Hz); 2.17 (*s*, 3H); 2.03 (*s*, 3H); 1.84 (*m*, 1H); 1.66 (*m*, 1H); 0.85 (*t*, 3H, $J = 7.5$ Hz). ^{13}C NMR (150 MHz, $CDCl_3$): δ (ppm): 165.6 (C); 160.1 (C); 151.7 (C); 137.4 (C); 137.3 (C); 135.0 (C); 129.3 (CH); 128.3 (CH); 127.6 (CH); 124.4 (C); 124.0 (CH); 123.9 (CH); 123.8 (C); 114.6 (CH); 80.4 (CH_2); 53.9 (CH_3); 47.8 (CH); 43.1 (CH); 24.6 (CH_2); 21.0 (CH_3); 18.2 (CH_3);11.7 (CH_3).

(Z)-methyl-3-((3*R*,4*R*)-3-ethyl-4-(2-fluorophenyl)-5-nitropentan-2-ylidene)-2-oxoindoline-1-carboxylate (product **126**, Table 2.4—entry 7)

126

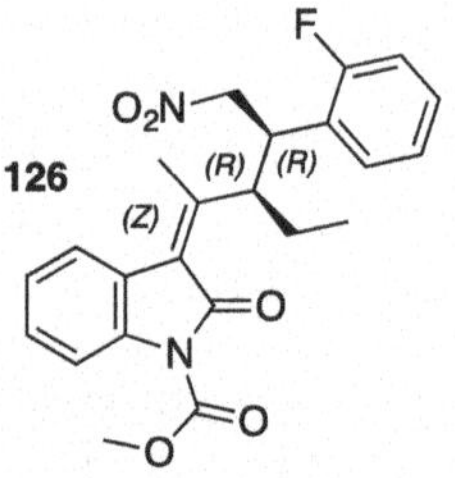

The title compound was obtained as single diastereoisomer. After purification by flash column chromatography (hexane/ethyl acetate = 8/2) (Z)-**126** was obtained in 66% yield and >99% ee. HPLC analysis on a AD-H column: hexane/*i*-PrOH 90/10, flow rate 0.5 mL/min, $\lambda = 214$ nm: $\tau_{major} = 16.19$ min. $[\alpha]_D^{20}$ +121.0 (*c* 1.00, $CHCl_3$). HRMS-ESI (+): calculated for $C_{23}H_{23}FNaN_2O_5$ 449.1483, found 449.1472 $[M+Na]^+$. ^1H NMR (600 MHz, $CDCl_3$): δ (ppm): 7.92 (*d*, 1H, $J = 8.1$ Hz); 7.39

(*d*, 1H, *J* = 7.8 Hz); 7.28 (*m*, 2H); 7.09 (*m*, 2H); 6.97 (*ddd*, 1H, J_1 = 7. 6 Hz, J_2 = 1.1 Hz); 6.89 (*ddd*, 2H, J_1 = 8.2 Hz, J_2 = 1.2 Hz); 5.31 (*ddd*, 1H, J_1 = 10.5 Hz, J_2 = 3.9 Hz); 4.90 (*dd*, 1H, J_1 = 12.8 Hz, J_2 = 4.7 Hz); 4.75 (*dd*, 1H, J_1 = 12.8 Hz, J_2 = 10.2 Hz); 4.09 (*ddd*, 1H, J_1 = 10.3 Hz, J_2 = 4.6 Hz); 4.05 (*s*, 3H); 2.08 (*s*, 3H); 1.87 (*m*, 1H); 1.70 (*m*, 1H); 0.86 (*t*, 3H, *J* = 7.4 Hz). ^{13}C NMR (150 MHz, $CDCl_3$): δ (ppm): 165.4 (C); 160.3 (C, *J* = 246.5 Hz); 158.1 (C); 150.7 (C); 136.4 (C); 129.4 (CH,, *J* = 8.3 Hz);129.1 (CH, *J* = 3.5 Hz); 128.5 (CH); 125.2 (C, *J* = 14.0 Hz); 123.6 (C); 124.4 (CH, *J* = 3.4 Hz); 123.1 (CH); 122.9 (CH); 122.6 (C); 115.6 (CH, *J* = 23.1 Hz); 114.5 (CH); 78.9 (CH_2); 53.9 (CH_3); 43.0 (CH); 24.6 (CH_2); 18.0 (CH) 17.9 (CH_3); 11.6 (CH_3).

(Z)-methyl-3-((3*R*,4*R*)-4-(4-bromophenyl)-3-ethyl-5-nitropentan-2-ylidene)-2-oxoindoline-1-carboxylate (product **127**, Table 2.4—entry 8)

127

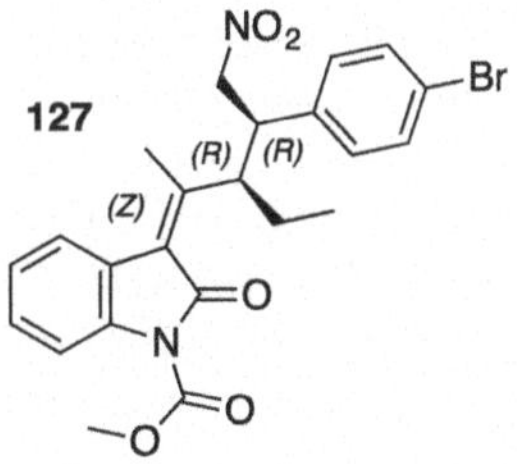

The title compound was obtained as single diastereoisomer. After purification by flash column chromatography (hexane/ethyl acetate = 8/2) (Z)-**127** was obtained in 62% yield and >99% ee. HPLC analysis on a AD-H column: hexane/*i*-PrOH 90/10, flow rate 0.5 mL/min, λ = 214 nm: τ_{major} = 19.38 min. $[\alpha]_D^{20}$ +146.0 (*c* 1.00, $CHCl_3$). HRMS-ESI (+): calculated for $C_{23}H_{23}BrNaN_2O_5$ 509.0683, found 509.0684 $[M+Na]^+$. ^{1}H NMR (600 MHz, $CDCl_3$): δ (ppm): 7.95 (*d*, 1H, *J* = 8.2 Hz); 7.39 (*d*, 1H, *J* = 7.9 Hz); 7.30 (*m*, 3H); 7.11 (*m*, 3H); 5.29 (*ddd*, 1H, J_1 = 10.6 Hz, J_2 = 4.0 Hz); 4.87 (*dd*, 1H, J_1 = 12.6 Hz, J_2 = 4.4 Hz); 4.61 (*dd*, 1H, J_1 = 12.5 Hz, J_2 = 10.7 Hz); 4.06 (*s*, 3H); 3.63 (*ddd*, 1H, J_1 = 10.6 Hz, J_2 = 4.4 Hz); 2.04 (*s*, 3H); 1.83 (*m*, 1H); 1.66 (*m*, 1H); 0.85 (*t*, 3H, *J* = 7.6 Hz). ^{13}C NMR (150 MHz, $CDCl_3$): δ (ppm): 164.6 (C); 157.8 (C); 150.5 (C); 136.4 (C); 136.3 (C); 130.8 (CH); 128.5 (CH); 127.7 (CH); 123.8 (C); 123.2 (CH); 122.9 (CH); 122.5 (C); 120.9 (CH); 113.7 (CH); 79.0 (CH_2); 53.0 (CH_3); 46.5 (CH); 41.8 (CH); 23.5 (CH_2); 17.0 (CH_3); 10.7 (CH_3).

(Z)-methyl-3-((4S)-3-ethyl-6-methyl-4-(nitromethyl)heptan-2-ylidene)-2-oxoindoline-1-carboxylate (product **128**, Table 2.4—entry 9)

128

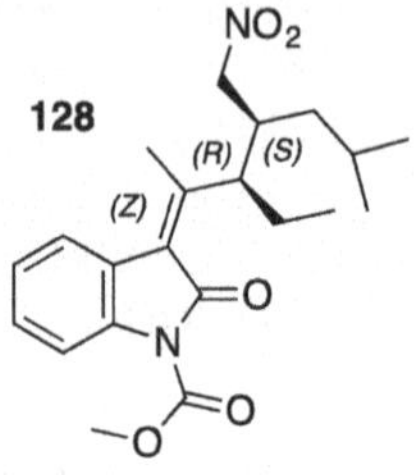

The title compound was obtained as single diastereoisomer. After purification by flash column chromatography (hexane/ethyl acetate = 85/15) (Z)-**128** was obtained in 20% yield and 99% ee. HPLC analysis on a AD-H column: hexane/*i*-PrOH 95/5, flow rate 0.5 mL/min, $\lambda = 254$ nm: $\tau_{major} = 14.06$ min.; $\tau_{minor} = 15.90$ min. $[\alpha]_D^{20}$ +11.0 (*c* 0.25, $CHCl_3$). HRMS-ESI (+): calculated for $C_{21}H_{28}NaN_2O_5$ 411.1890, found 411.1891 $[M+Na]^+$. ^1H NMR (600 MHz, $CDCl_3$): δ (ppm): 8.03 (*d*, 1H, *J* = 8.1 Hz); 7.65 (*d*, 1H, *J* = 8.1 Hz); 7.37 (*m*, 1H); 7.23 (*m*, 1H); 4.79 (*ddd*, 1H, J_1 = 10.8 Hz, J_2 = 4.0 Hz); 4.51 (*dd*, 1H, J_1 = 13.6 Hz, J_2 = 6.0 Hz); 4.40 (*dd*, 1H, J_1 = 13.4 Hz, J_2 = 4.4 Hz); 4.04 (*s*, 3H); 2.53 (*m*, 1H); 2.26 (*s*, 3H); 1.73 (*m*, 1H); 1.49 (*m*, 1H); 1.32 (*m*, 2H); 1.01 (*m*, 1H); 0.83 (*m*, 6H); 0.77 (*t*, 3H, *J* = 7.3 Hz). ^{13}C NMR (150 MHz, $CDCl_3$): δ (ppm): 165.5, 161.0, 151.8, 137.6, 128.6, 125.0, 124.2, 124.0, 123.9, 114.8, 78.5, 53.9, 45.0, 40.6, 38.0, 25.2, 23.8, 23.1, 21.3, 17.7, 11.7.

(Z)-methyl-3-((4R,5R)-4-ethyl-6-nitro-5-phenylhexan-3-ylidene)-2-oxoindoline-1-carboxylate (product **129**, Table 2.4—entry 10)

129

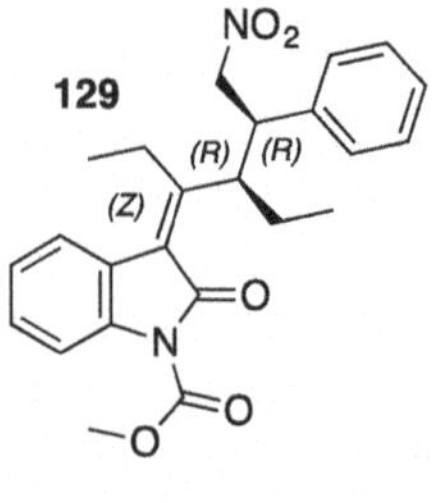

The title compound was obtained as single diastereosiomer. After purification by flash column chromatography (hexane/ethyl acetate = 8/2) (Z)-**129** was obtained in 44% yield and 99% ee. HPLC analysis on a AD-H column: hexane/*i*-PrOH 95/5, flow rate 0.5 mL/min, $\lambda = 254$ nm: $\tau_{major} = 19.11$ min; $\tau_{minor} = 24.48$ min. $[\alpha]_D^{20}$ +185.8 (*c* 1.00, $CHCl_3$). HRMS-ESI (+): calculated for $C_{24}H_{26}NaN_2O_5$ 445.1734, found 445.1732 $[M+Na]^+$. ^1H NMR (400 MHz, $CDCl_3$): δ (ppm): 7.92 (*d*, 1H, *J* = 8.5 Hz); 7.27 (*m*, 3H); 7.11 (*m*, 5H); 5.37 (*ddd*, 1H, J_1 = 10.6 Hz, J_2 = 3.6 Hz); 4.88 (*dd*, 1H, J_1 = 12.4 Hz, J_2 = 4.5 Hz); 4.65 (*dd*, 1H, J_1 = 12.6 Hz, J_2 = 10.2 Hz); 4.06 (*s*, 3H); 3.69 (*ddd*, 1H, J_1 = 10.3 Hz, J_2 = 4.5 Hz); 2.50 (*m*, 1H); 2.33 (*m*, 1H);

1.87 (*m*, 1H); 1.76 (*m*, 1H); 1.11 (*t*, 3H, $J = 7.7$ Hz); 0.94 (*t*, 3H, $J = 7.7$ Hz). ^{13}C NMR (100 MHz, $CDCl_3$): δ (ppm): 166.1, 165.7, 151.7, 138.4, 137.3, 128.5, 128.4, 127.8, 127.7, 124.9, 124.2, 123.7, 122.4, 114.5, 80.6, 53.9, 48.2, 43.6, 24.2, 24.0, 12.8, 11.9.

(Z)-methyl-3-((*R*)-5-methyl-3-((*R*)-2-nitro-1-phenylethyl)hex-5-en-2-ylidene)-2-oxoindoline-1-carboxylate (product **130**, Table 2.4—entry 11)

130

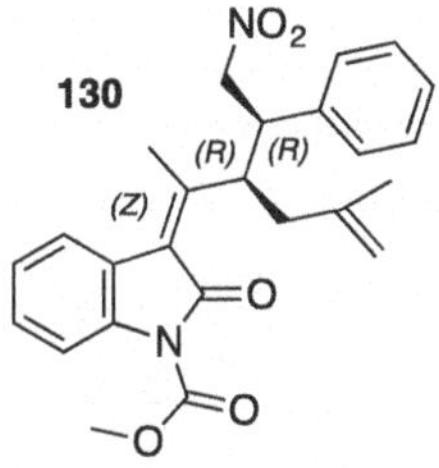

The title compound was obtained as single diastereoisomer. After purification by flash column chromatography (hexane/ethyl acetate = 8/2) (Z)-**130** was obtained in 81% yield and >99% ee. HPLC analysis on a cellulose-2 column: hexane/*i*-PrOH 90/10, flow rate 0.5 mL/min, λ = 214 nm: $\tau_{major} = 49.53$ min. $[\alpha]_D^{20}$ +135.2 (*c* 1.00, $CHCl_3$). HRMS-ESI (+): calculated for $C_{25}H_{26}NaN_2O_5$ 457.1734, found 457.1730 $[M+Na]^+$. ^{1}H NMR (400 MHz, $CDCl_3$): δ (ppm): 7.90 (*d*, 1H, $J = 8.4$ Hz); 7.34 (*d*, 1H, $J = 8.4$ Hz); 7.18 (*m*, 7H); 5.54 (*m*, 1H); 4.95 (*dd*, 1H, $J_1 = 12.5$ Hz, $J_2 = 4.4$ Hz); 4.73 (*m*, 3H); 4.05 (*s*, 3H); 3.70 (*ddd*, 1H, $J_1 = 10.0$ Hz, $J_2 = 4.1$ Hz); 2.47 (*m*, 2H); 2.01 (*s*, 3H); 1.74 (*s*, 3H). ^{13}C NMR (100 MHz, $CDCl_3$): δ (ppm): 164.5, 158.3, 150.6, 141.6, 136.8, 136.4, 127.5, 127.4, 126.9, 123.2, 123.0, 122.9, 122.7, 113.5, 112.2, 78.6, 52.9, 47.5, 39.8, 39.2, 21.2, 17.8.

(Z)-methyl 3-((3*R*,4*R*)-3-benzyl-5-nitro-4-phenylpentan-2-ylidene)-2-oxoindoline-1-carboxylate (product **131**, Table 2.4—entry 12)

131

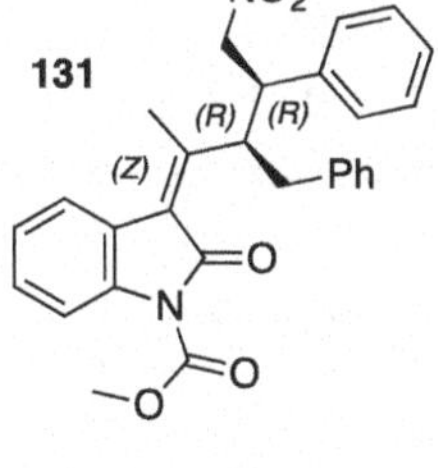

The title compound was obtained as single diastereoisomer. After purification by flash column chromatography (hexane/ethyl acetate = 8/2) (Z)-**131** was obtained in 50% yield and 96% ee. The ee was determined by HPLC analysis on a AD-H column: hexane/*i*-PrOH 95/5, flow rate 0.5 mL/min, λ = 254 nm: $\tau_{major} = 43.36$ min; $\tau_{minor} = 38.83$ min. $[\alpha]_D^{20}$ +123.7 (*c* 1.00, $CHCl_3$). HRMS-ESI (+): calculated for

$C_{28}H_{26}NaN_2O_5$ 493.1734, found 493.1735 $[M+Na]^+$. ^{1}H NMR (600 MHz, $CDCl_3$): δ (ppm): 7.84 (*d*, 1H, $J = 8.5$ Hz); 7.18 (*m*, 12H); 7.02 (*dd*, 1H, $J = 8.5$ Hz); 5.76 (*m*, 1H); 4.83 (*dd*, 1H, $J_1 = 12.4$ Hz, $J_2 = 4.2$ Hz); 4.65 (*dd*, 1H, $J_1 = 12.3$ Hz, $J_2 = 10.7$ Hz); 4.05 (*s*, 3H); 3.82 (*dd*, 1H, $J_1 = 10.2$ Hz, $J_2 = 4.2$ Hz); 3.17 (*dd*, 1H, $J_1 = 14.1$ Hz, $J_2 = 6.0$ Hz); 2.92 (*dd*, 1H, $J_1 = 14.1$ Hz, $J_2 = 9.0$ Hz); 1.99 (*s*, 3H). ^{13}C NMR (150 MHz, $CDCl_3$): δ (ppm): 165.3, 158.5, 151.5, 137.9, 137.3, 128.7, 128.6, 128.5, 128.4, 127.9, 127.8, 126.7, 124.3, 123.9, 123.7, 123.5, 114.5, 80.0, 53.8, 48.0, 42.9, 38.5, 18.9.

(*S*,*Z*)-methyl-5-chloro-3-(1-ethoxy-2-(nitromethyl)-1-oxo-2-phenylheptan-4-ylidene)-2-oxoindoline-1-carboxylate (product **104**, Scheme 2.3)

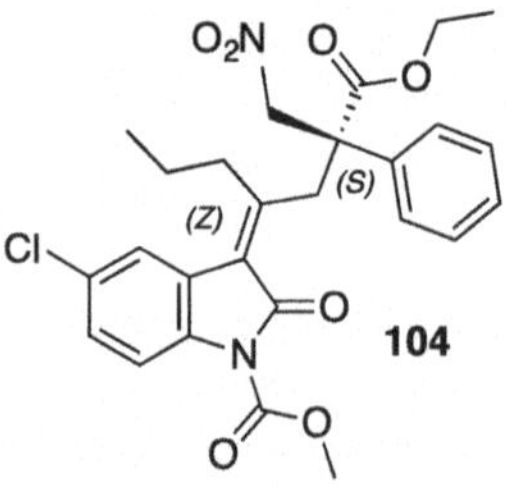

104

The title compound was obtained as single diastereoisomer. After purification by flash column chromatography (hexane/ethyl acetate = 8/2) (*Z*)-**104** was obtained in 87% yield and >99% ee. HPLC analysis on a AD-H column: hexane/*i*-PrOH 95/5, flow rate 1.0 mL/min, $\lambda = 254$ nm: $\tau_{major} = 10.66$ min; $\tau_{minor} = 6.05$ min. $[\alpha]_D^{20}$ +66.1 (*c* 1.00, $CHCl_3$). HRMS-ESI (+): calculated for $C_{26}H_{27}ClNaN_2O_7$ 537.1399, found 537.1401 $[M+Na]^+$. ^{1}H NMR (400 MHz, $CDCl_3$): δ (ppm): 7.89 (*d*, 1H, $J = 8.9$ Hz); 7.39 (*bs*, 1H); 7.34 (*m*, 4H); 7.27 (*m*, 2H); 5.45 (*d*, 1H, $J = 15.8$ Hz); 5.16 (*d*, 1H, $J = 15.8$ Hz); 4.30 (*m*, 2H); 4.10 (*d*, 1H, $J = 13.5$ Hz); 4.04 (*s*, 3H); 3.81 (*d*, 1H, $J = 13.5$ Hz); 2.57 (*m*, 1H); 2.49 (*m*, 1H); 1.53 (*m*, 2H); 1.03 (*t*, 3H, $J = 7.2$ Hz). ^{13}C NMR (100 MHz, $CDCl_3$): δ (ppm): 171.2, 164.6, 163.0, 150.3, 137.4, 135.0, 128.9, 127.8, 127.4, 127.1, 125.2, 123.2, 122.8, 122.7, 114.8, 78.0, 61.2, 54.2, 53.0, 39.1, 37.0, 28.7, 19.4, 13.3, 12.8.

(*Z*)-methyl-3-((3*S*,4*S*)-5-ethoxy-3-ethyl-4-(nitromethyl)-5-oxo-4-phenylpentan-2-ylidene)-2-oxoindoline-1-carboxylate (product **105**, Scheme 2.3)

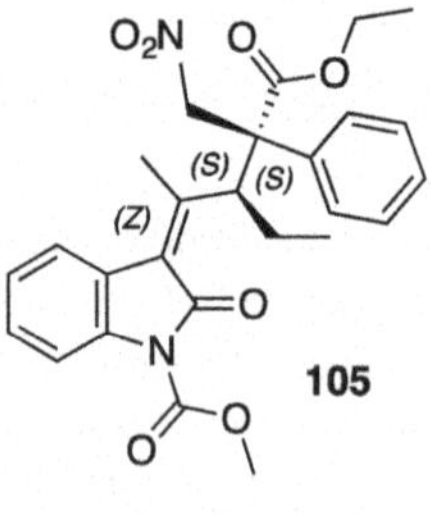

105

The title compound was obtained as single diastereoisomer. After purification by flash column chromatography (hexane/ethyl acetate = 8/2) (Z)-**105** was obtained in 50% yield and >99% ee. HPLC analysis on a AD-H column: hexane/*i*-PrOH 90/10, flow rate 0.5 mL/min, $\lambda = 214$ nm: $\tau_{major} = 15.16$ min. $[\alpha]_D^{20}$ +13.0 (*c* 1.00, $CHCl_3$). HRMS-ESI (+): calculated for $C_{26}H_{28}NaN_2O_7$ 503.1789, found 503.1787 $[M+Na]^+$. ^1H NMR (600 MHz, $CDCl_3$): δ (ppm): 8.00 (*d*, 1H, $J = 8.2$ Hz); 7.58 (*d*, 1H, $J = 7.3$ Hz); 7.45 (*d*, 1H, $J = 7.9$ Hz); 7.34 (*m*, 5H); 7.16 (*m*, 1H); 5.76 (*d*, 1H, $J = 15.5$ Hz); 5.49 (*dd*, 1H, $J_1 = 11.7$ Hz, $J_2 = 3.2$ Hz); 5.19 (*d*, 1H, $J = 15.4$ Hz); 4.36 (*m*, 2H); 4.08 (*s*, 3H); 1.79 (*m*, 1H); 1.70 (*m*, 1H); 1.50 (*s*, 3H); 1.34 (*t*, 3H, $J = 7.0$ Hz); 0.71 (*t*, 3H, $J = 7.2$ Hz). ^{13}C NMR (150 MHz, $CDCl_3$): δ (ppm): 171.2, 164.6, 163.0, 150.3, 137.4, 135.0, 128.9, 127.8, 127.4, 127.1, 125.2, 123.2, 122.8, 122.7, 114.8, 78.0, 61.2, 54.2, 53.0, 39.1, 37.0, 28.7, 19.4, 13.3, 12.8.

(*R*,*Z*)-methyl-5-chloro-3-(1-nitro-2-phenylheptan-4-ylidene-3,3,5,5-*d*$_4$)-2-oxoindoline-1-carboxylate (product **91**, Scheme 2.1).

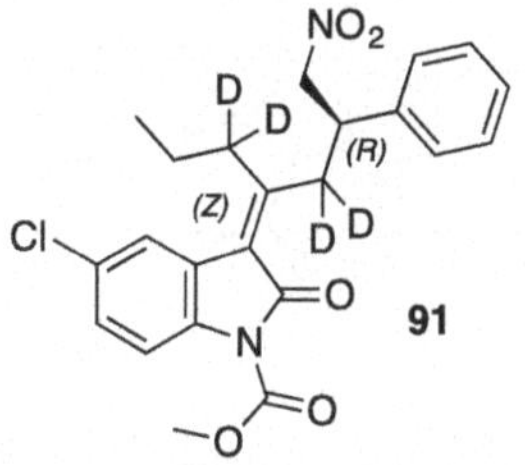

91

The title compound was obtained as an amorphous solid following the general procedure and was treated exactly like compound (Z)-**91**. HRMS-ESI (+): calculated for $C_{23}H_{19}D_4ClNaN_2O_5$ 469.1547, found 469.1560 $[M+Na]^+$. ^1H NMR (600 MHz, $CDCl_3$): δ (ppm): 7.95 (*d*, 1H, $J = 8.6$ Hz); 7.40 (*d*, 1H, $J = 2.0$ Hz); 7.35–7.25 (*m*, 6H); 4.79 (*dd*, 1H, $J_1 = 12.9$ Hz, $J_2 = 9.9$ Hz); 4.71 (*dd*, 1H, $J_1 = 12.9$ Hz, $J_2 = 5.5$ Hz); 4.07 (*s*, 3H); 3.84 (*m*, 1H); 1.53 (*m*, 2H); 1.05 (*t*, 3H, $J = 7.5$ Hz). ^{13}C NMR (150 MHz, $CDCl_3$): δ (ppm): 165.1, 163.1, 151.4, 139.2, 136.3, 129.9, 129.0, 128.5, 128.0, 127.5, 124.1, 123.4, 122.6, 116.1, 79.4, 54.1, 43.5, 37.9, 31.5, 20.1, 14.3.

(*R*,*Z*)-methyl-5-chloro-3-(1-nitro-2-phenylheptan-4-ylidene-3,3-*d*$_2$)-2-oxoindoline-1-carboxylate (product **90**, Scheme 2.1).

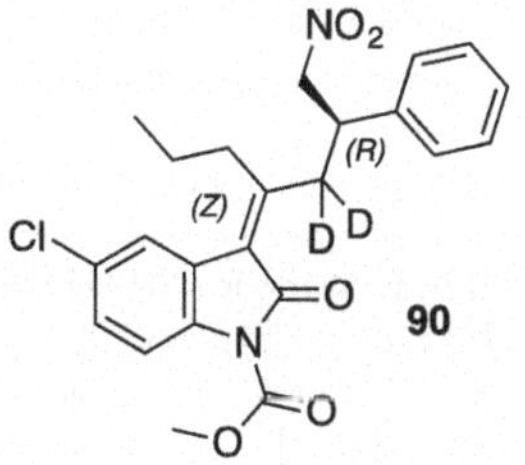

90

The title compound was obtained as an amorphous solid following the general procedure and was treated exactly like compound (*Z*)-**90**. HRMS-ESI (+): calculated for $C_{23}H_{21}D_2ClNaN_2O_5$ 467.1421, found 467.1416 $[M+Na]^+$. ^{1}H NMR (600 MHz, $CDCl_3$): δ (ppm): 7.95 (*d*, 1H, $J = 8.9$ Hz); 7.40 (*bs*, 1H); 7.35–7.25 (*m*, 6H); 4.79 (*dd*, 1H, $J_1 = 12.9$ Hz, $J_2 = 9.9$ Hz); 4.71 (*dd*, 1H, $J_1 = 12.9$ Hz, $J_2 = 5.6$ Hz); 4.07 (*s*, 3H); 4.03 (*m*, 0.8H); 3.85 (*m*, 1H); 2.83 (*m*, 0.5H); 2.64 (*m*, 1H); 2.23 (*m*, 1H); 1.54 (*m*, 2H); 1.05 (*t*, 3H, $J = 7.2$ Hz). ^{13}C NMR (150 MHz, $CDCl_3$): δ (ppm): 165.1, 163.3, 151.4, 139.2, 136.3, 129.9, 129.0, 128.5, 128.0, 127.5, 124.1, 123.4, 122.7, 116.1, 79.3, 54.1, 43.5, 38.4, 31.6, 20.3, 14.3.

(Z)-methyl-5-chloro-3-((3*R*,4*R*)-3-ethyl-5-nitro-4-phenylpentan-2-ylidene-1,1,1-*d*$_3$)-2-oxoindoline-1-carboxylate (product **94**, Scheme **2.1**).

94

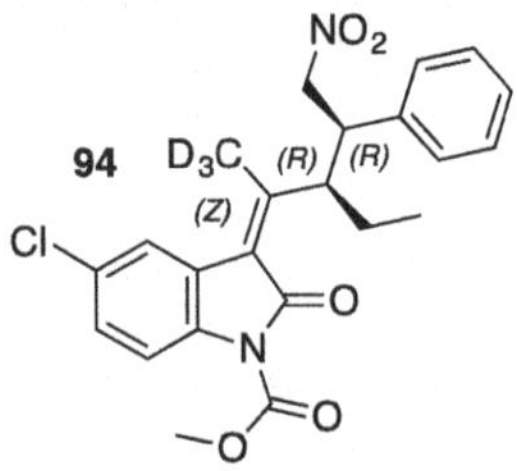

The title compound was obtained as an amorphous solid following the general procedure and was treated exactly like compound (Z)-**94**. HRMS-ESI (+): calculated for $C_{23}H_{20}D_3ClNaN_2O_5$ 468.1484, found 468.1480 $[M+Na]^+$. ^{1}H NMR (600 MHz, $CDCl_3$): δ (ppm): 7.87 (*d*, 1H, $J = 8.9$ Hz); 7.30 (*bs*, 1H); 7.23 (*dd*, 1H, $J_1 = 12.9$ Hz, $J_2 = 2.1$ Hz); 7.20–7.10 (*m*, 4H); 5.28 (*m*, 0.9H); 4.88 (*dd*, 1H, $J_1 = 12.6$ Hz, $J_2 = 4.5$ Hz); 4.66 (*dd*, 1H, $J_1 = 12.6$ Hz, $J_2 = 10.4$ Hz); 4.05 (*s*, 3H); 3.65 (*m*, 1H); 2.02 (*m*, 1.5H); 1.87 (*m*, 1H); 1.68 (*m*, 1H); 0.85 (*t*, 3H, $J = 7.5$ Hz). ^{13}C NMR (150 MHz, $CDCl_3$): δ (ppm): 165.0, 161.8, 151.5, 138.0, 135.7, 129.6, 128.6, 128.1, 127.9, 127.7, 124.9, 123.9, 115.7, 80.1, 54.0, 48.0, 43.4, 24.5, 18.0, 11.7.

References

1. Fuson RC (1935) Chem. Rev 1
2. Pansare SV, Paul EK (2011) Chem. Eur. J 8770
3. Hepburn HB, Dell'amico L, Melchiorre P (2016) Chem. Rec. 1787
4. Curti C, Rassu G, Zambrano V, Pinna L, Pelosi G, Sartori A, Battistini L, Zanardi F, Casiraghi G (2012) Angew. Chem. Int. Ed. 6200
5. Rassu G, Zambrano V, Pinna L, Curti C, Battistini L, Sartori A, Pelosi G, Zanardi F, Casiraghi G (2013) Adv. Synth. Catal
6. Chen Q, Wang G, Jiang X, Xu Z, Lin L, Wang R (2014) Org. Lett. 1394
7. Still WC, Kahn M, Mitra AJ (1978) J. Org. Chem 43:2923
8. Vakulya B, Varga S, Csámpai A, Soós T (2005) Org. Lett. 10:1967
9. Trost BM, Cramer N, Silverman SM (2007) J. Am. Chem. Soc 129:12396

Chapter 3
Targeting the Remote Control of Axial Chirality in *N*-(2-*tert*-butylphenyl)Succinimides via a Desymmetrization Strategy

Abstract There are molecules possessing stereogenic elements that are not chiral. These compounds are termed "meso" and do not manifest chirality because they possess a symmetry element of the second order (i.e. a symmetry plane)[2] that makes their mirror images superimposable (Fig. 3.1).

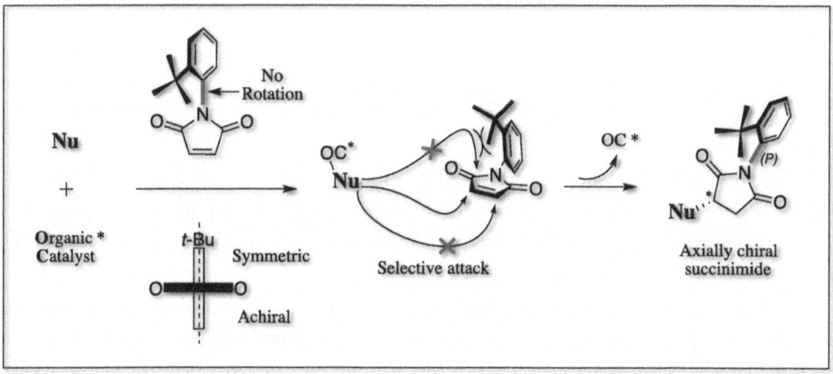

Parts of this chapter were adapted from:

- "N. Di Iorio, P. Righi, A. Mazzanti, M. Mancinelli, A. Ciogli, G. Bencivenni *J. Am. Chem. Soc.*, **2014**, 10,250" with permission from JOURNAL of AMERICAL CHEMICAL SOCIETY, copyright 2014 American Chemical Society
- "N. Di Iorio, F. Champavert, A. Erice, P. Righi, A. Mazzanti, G. Bencivenni *Tetrahedron (invited paper)*, **2016**, 5191—special issue: Methods for controlling axial chirality" with permission from ELSEVIER, copyright 2016

Published as: References [1–3].

N. Di Iorio, *New Organocatalytic Strategies for the Selective Synthesis of Centrally and Axially Chiral Molecules*, Springer Theses,
https://doi.org/10.1007/978-3-319-74914-3_3

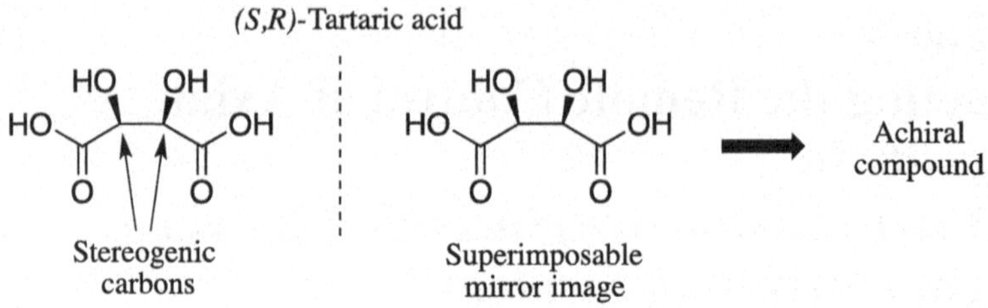

Fig. 3.1 Meso form of tartaric acid

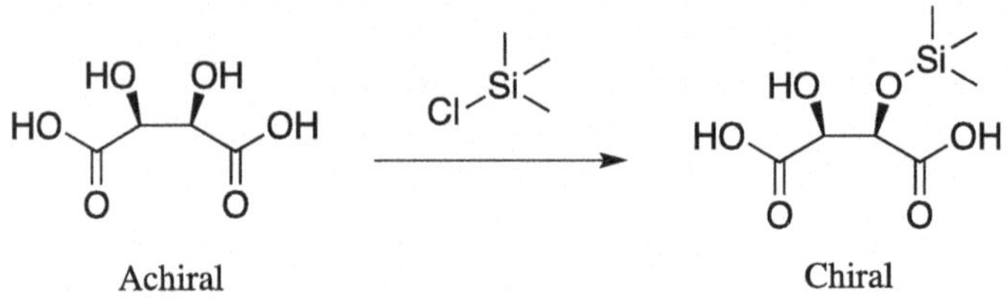

Fig. 3.2 Desymmetrization of meso tartaric acid

Original content can be found at http://pubs.acs.org/doi/abs/10.1021/ja505610k and http://www.sciencedirect.com/science/article/pii/S0040402016301144.

3.1 Desymmetrization as a Tool for Asymmetric Synthesis

There are molecules possessing stereogenic elements that are not chiral. These compounds are termed "meso" and do not manifest chirality because they possess a symmetry element of the second order (i.e. a symmetry plane) [4] that makes their mirror images superimposable (Fig. 3.1).

The *S,R* diastereoisomer of tartaric acid is a meso form and is not chiral, but breaking through its symmetry, for example by protection of one OH group, would reveal the central chirality of the two stereocenters and the resulting molecule would be chiral itself (Fig. 3.2).

Stereoselective derivatization of meso compounds (but also of prochiral compounds) is a very useful method for the synthesis of chiral molecules[1] and the concept of desymmetrization has recently found large employment in asymmetric catalysis. One remarkable example was reported by Dixon and consists in a stereoselective intramolecular Michael addition (Reaction 3.1) [7].

Combining enamine and H-bonding catalysis, the authors were able to forge two new stereocenters and reveal the chirality of the prochiral spiro carbon with high yield and selectivity. In this project, we apply the same desymmetrization strategy to axial chirality on ortho-substituted *N*-aryl maleimides. It is known that when the

[1]For further reading on desymmetrization see: [5, 6].

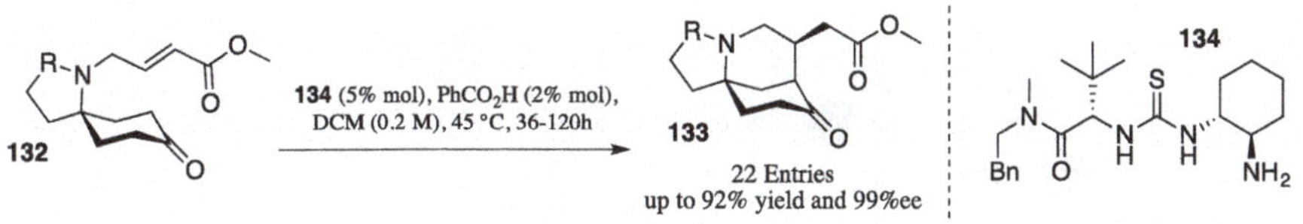

Reaction 3.1 Desymmetrization reaction of prochiral cyclohexanones

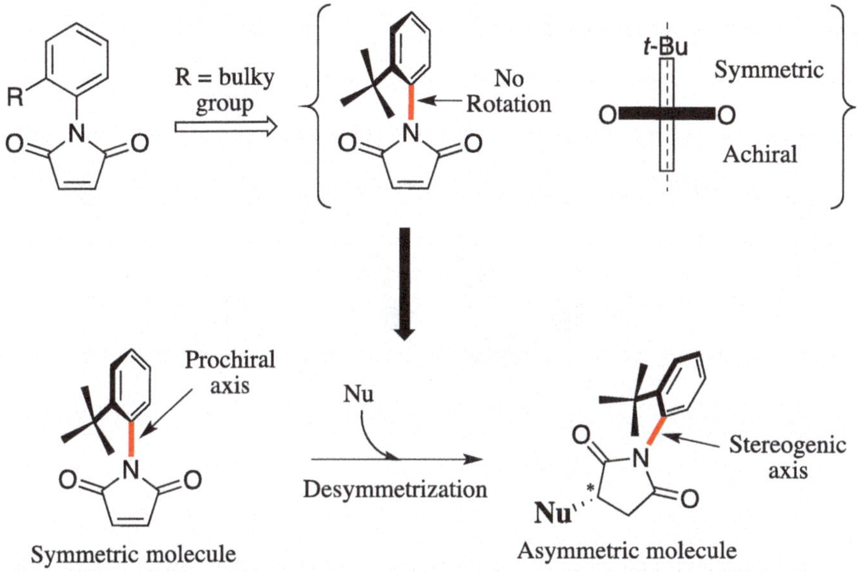

Scheme 3.1 Our desymmetrization strategy

ortho substituent in these compounds is bulky enough [8], the rotation of the imide N–C bond freezes generating a prochiral axis (Scheme 3.1).

The aim of this work is to see if it is possible to break through the symmetry of the molecule and transmit the chiral information of a catalyst to the newly formed stereogenic axis from a remote site of the molecule and to develop a general protocol working for nucleophiles of various nature.

3.2 Results and Discussion

We began our investigation using 3-alkylcyclohexanones as nucleophiles because, upon activation with a primary amine [9], they are known to generate a nucleophilic dienamine intermediate. We found it to be reactive towards the maleimide generating products with two adjacent stereocenters, three and four bonds away from the

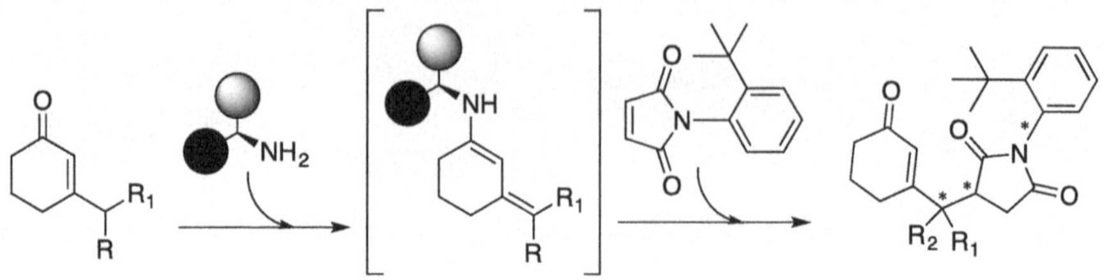

Fig. 3.3 Products of the desymmetrization

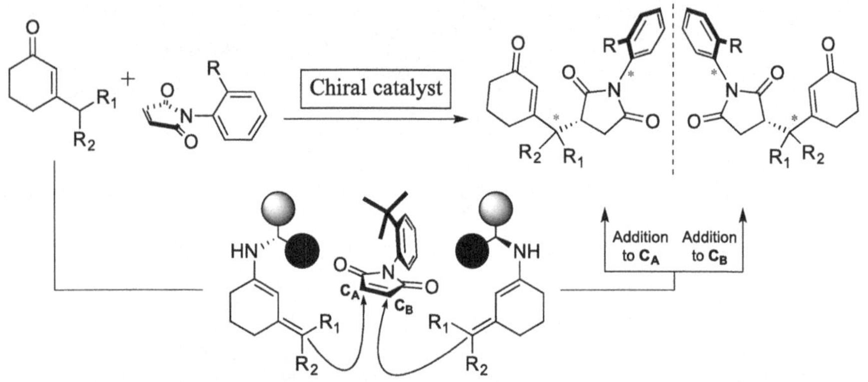

Scheme 3.2 Origin of the enantioselectivity

activated carbonyl, and an even more distant stereogenic axis, seven bonds away from the active site (Fig. 3.3).[2]

For this strategy to be successful we needed a catalyst that would direct the nucleophile towards just one of the two faces of the maleimide (most likely the one opposite to the *t*-butyl group), but also towards only one of its two electrophilic carbons in order to obtain a single enantiomer of the product (Scheme 3.2).

So we started by screening some primary amines in combination with some acidic cocatalysts for the reaction of model substrates **135** and **136** (Table 3.1).

We first attempted the reaction with DHQDA catalyst **139** and observed a moderate reactivity for the formation of two optically pure diastereoisomers (**137a** and **137b**) in a 70:30 ratio. We determined the absolute configuration of major diastereoisomer **137a** to be *P,R,R*,[3] and consequent NOE experiments allowed us to assign a *P,R,S* absolute configuration to minor diastereoisomer **137b** (Fig. 3.4).

These results showed that we had complete control over the stereogenic axis because the nucleophilic attack is completely selective and takes place only towards

---

[2]We have already talked in Sect. 2.1 about the issues arising from vinylogous reactivity where usually the stereogenic carbons are three or four bonds away from the activated site of the molecule therefore the challenge of forging a stereogenic element seven bonds away is even more formidable.

[3]The absolute configuration was determined via single crystal XRD of the corresponding brominated product **149a** and assigned by analogy to **137a**.

Table 3.1 Screening of the reaction conditions

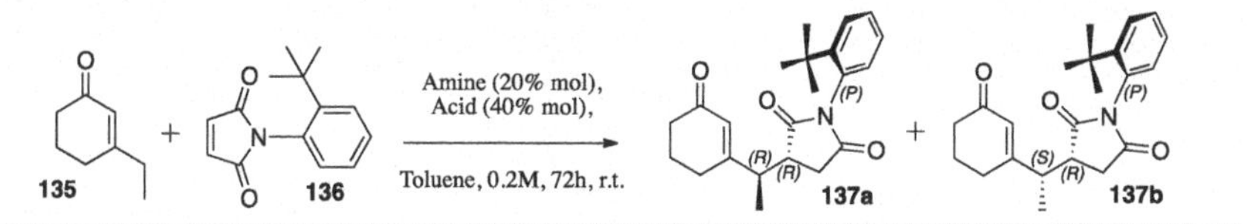

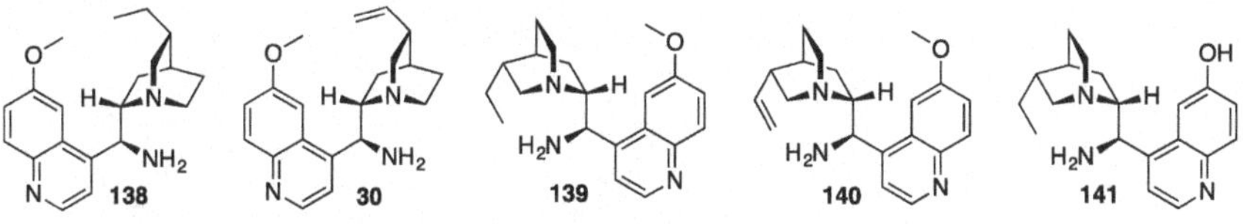

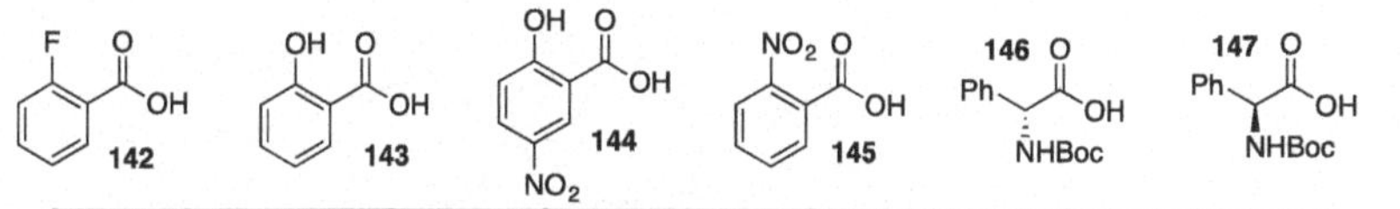

Entry	Catalyst	Acid	Yield (137a+137b)	d.r. (137a:137b)	ee% (137a/137b)
1	**139**	**142**	43	70:30	>99/>99
2	**138**	**142**	30	70:30	>-99/>-99
3	**141**	**142**	33	70:30	>-99/>-99
4	**139**	**143**	30	70:30	>99/>99
5	**139**	**144**	25	70:30	>99/>99
6	**139**	**145**	42	70:30	>99/>99
7[a]	**139**	**142**	58	67:33	>99/>99
8[a]	**139**	**146**	51	65:35	>99/>99
9[a]	**139**	**147**	60	65:35	>99/>99
10[a]	**140**	**146**	70	70:30	>99/>99
11[a]	**140**	**147**	75	70:30	>99/>99
12[a]	**30**	**146**	43	70:30	>-99/>-99
13[a]	**30**	**147**	35	70:30	>-99/>-99
14[a]	**138**	**146**	40	65:35	>-99/>-99
15[a]	**138**	**147**	33	65:35	>-99/>-99

[a]Reaction performed with 2 equiv. of 135

one of its two electrophilic carbons as mentioned earlier and from one face of the maleimide (the one opposite to the *t*-butyl group) (Scheme 3.2). In an effort to improve the yield and the diastereoselectivity, we raised the equivalents of ketone from one to two and tried other cinchona-derived amines and acids. At the end of our screening, we observed the best result is obtained with QDA catalyst **140** and *N*-Boc-protected amino acid **147** (Table 3.1—entry 11) affording the product in 75% yield and again as a 70:30 mixture of diastereoisomers. Next, we set to evaluate the barrier

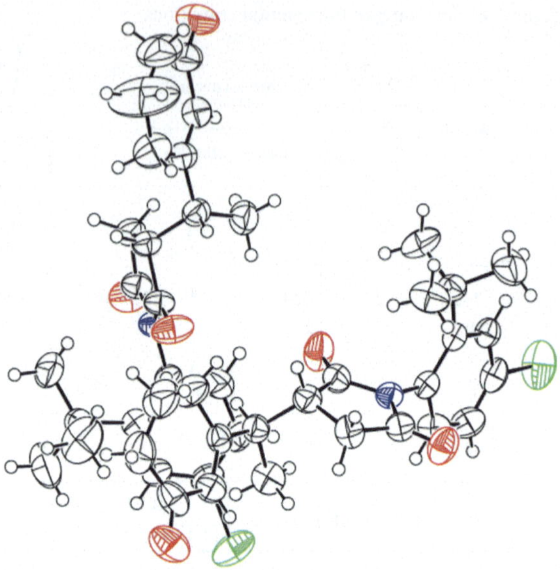

Fig. 3.4 XRD-derived structure of brominated product 149a

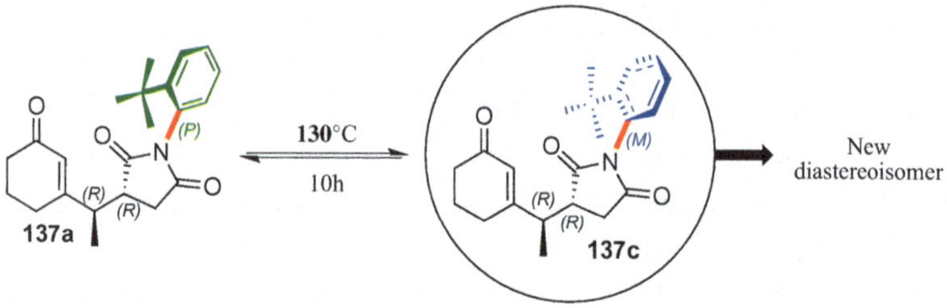

Scheme 3.3 Axial epimerization of the stereogenic axis

to rotation of the stereogenic axis of **137a**.[4] After heating it at 130 °C for 10 h, we observed a 62:38 equilibrium mixture of **137a** and **137c** that is a new diastereoisomer (never obtained at the end of the reactions) corresponding to the axial epimer and confirming once more that we had complete control of the remote stereogenic axis whose rotational barrier was found to be 31.9 kcal/mol (Scheme 3.3).[5]

At this point we decided to address the low diastereoselection. Having basically always the same d.r. with many catalytic systems, we suspected that epimerization was occurring at the exocyclic stereocenter so we isolated diastereoisomers **137a** and **137b** and left them separately under reaction conditions for 24 h (Scheme 3.4).

Treatment with the primary amine under reaction conditions afforded the usual 70:30 mixture of **137a**:**137b** regardless of the starting diastereoisomer meaning that epimerization does occur at the exocyclic stereocenter accounting for the low

[4]See the experimental section for details.

[5]A value of 31.9 kcal/mol corresponds roughly to a half life of epimerization of 1000 years.

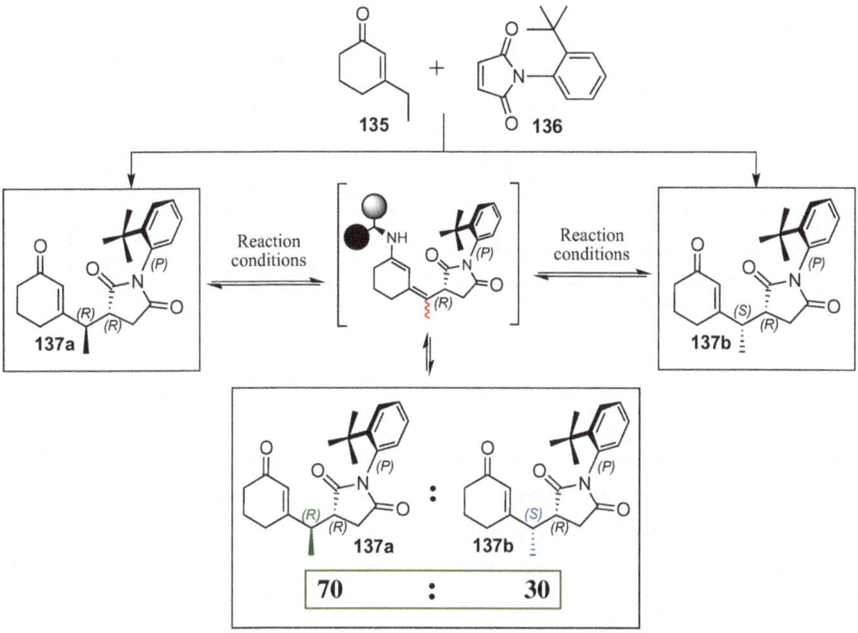

Scheme 3.4 Primary amine-catalyzed epimerization of the exocyclic stereocenter

selectivity. We could now move on to investigate the substrate scope using various enones and maleimides (Table 3.2).

The reaction showed steady d.r. values and almost complete enantioselectivity for every substrate, proceeded smoothly when halogen, phenyl and methoxy substituents were present (Table 3.2—entries 2–5) and tolerated also a protected amino group (Table 3.2—entries 6–7). Endocyclic substituted cyclohexenones afforded the corresponding succinimides in high yields (Table 3.2—entries 8–9), whereas exocyclic substituted enones gave the desired products in lower yields when reacted with maleimide **136** (Table 3.2—entries 10–12), but with more reactive 4-Br- and 5-NHCbz-substituted maleimides, the reactivity remained acceptable (Table 3.2—entries 13–14). We then expanded the scope by reacting some γ-disubstituted cyclohexenones in order to generate products with a quaternary center or stereocenter (Table 3.3).

Good yields and excellent diastereo- and enantioselectivity were obtained with *i*-propyl- and cyclopentyl-substituted enones (Table 3.3—entries 1–6). Non-symmetrically-substituted enones (Table 3.3—entries 7–9) gave once more very good enantioselectivity with slightly lower yields. Interestingly these substrates can not epimerize, in fact for the first time we observed a significantly different d.r. value with respect to 70:30. Also, NOESY experiments on major diastereoisomer **169a** showed an absolute configuration of *P,R,S* (opposite exocyclic carbon) in contrast with the trend observed for the products presented in Table 3.2. This most likely means that

Table 3.2 Substrate scope

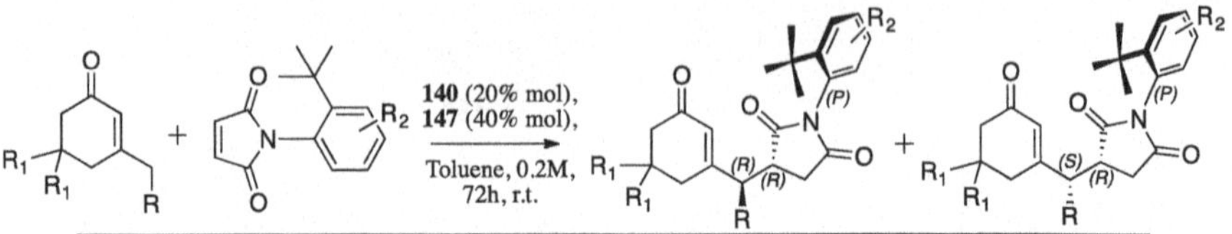

Entry	R/R[1]/R[2]	Product	Yield (a+b)	d.r. (a:b)	ee% (a/b)
1	Me/H/H	**148a+148b**	75	70 : 30	>99/>99
2	Me/H/4-Br	**149a+149b**	80	70 : 30	97/>99
3	Me/H/4-Cl	**150a+150b**	70	72 : 28	>99/>99
4	Me/H/4-Ph	**151a+151b**	75	70 : 30	96/95
5	Me/H/4-OMe	**152a+152b**	80	70 : 30	94[a]/-
6	Me/H/5-NHCbz	**153a+153b**	80	75 : 25	98/97
7	Me/H/5-NHTs	**154a+154b**	45	75 : 25	95[a]/-
8	Me/Me/H	**155a+155b**	60	70 : 30	96/95
9	Me/Me/4-Cl	**156a+156b**	60	65 : 35	95/95
10	Bn/H/H	**157a+157b**	50	70 : 30	99/97
11	*n*-Pr/H/H	**158a+158b**	37	64 : 36	97/94
12	Allyl/H/H	**159a+159b**	36	70 : 30	96/96
13	*n*-Pr/H/4-Br	**160a+160b**	76	70 : 30	99/98
14	Allyl/H/5-NHCbz	**161a+161b**	63	70 : 30	97/97

[a]Determined by ^{1}H-NMR using Pirkle alcohol

the major diastereoisomer afforded by the catalyst has always an *S* absolute configuration at the exocyclic carbon and we observe the opposite when epimerization occurs. Summing all these informations up and considering that the nucleophilic attack takes always place on the opposite face with respect to the *t*-Bu group, we could draw a reasonable transition state that accounts for the selectivity observed (Fig. 3.5).

The dienamine (also for γ-disubstituted cyclohexenones) has the thermodynamically favored *E* configuration on the terminal double bond and shows the *Re* face to the maleimide that is held in place by a crucial network of H-bonds. This way the only accessible carbon is $\mathbf{C_A}$ and the reaction proceeds selectively affording a single enantiomer.

At this point we checked the feasibility of this desymmetrization using other established nucleophiles like 1,3-dicarbonyl compounds, β-ketoesters, cyanoacetates and biologically relevant oxindoles (Scheme 3.5).

We quickly found that the desymmetrization is very general and works with the mentioned nucleophiles under Brønsted-base catalysis. DHQDA dimer **171** effi-

Table 3.3 Extension of the substrate scope

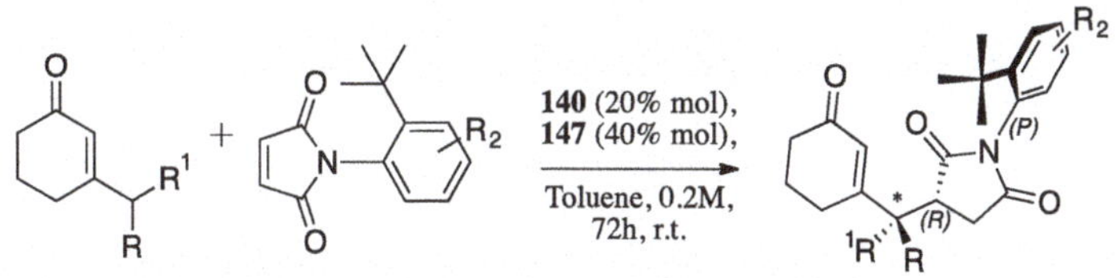

Entry	Product	R/R¹/R²	Yield	d.r. (a:b)	ee% (a/b)
1	**162**	Me/Me/H	83	-	96
2	**163**	Me/Me/4-Cl	75	-	99
3	**164**	Me/Me/4-Ph	68	-	97
4	**165**	Me/Me/4-OMe	74	-	98
5	**166**	Me/Me/5-NHCbz	81	-	97
6	**167**	-(CH2)4-/H	80	-	98
7	**168a+168b**	Me/Et/H	30 **(a+b)**	75:25	97/95
8	**169a+169b**	Me/Et/4-Br	35 **(a+b)**	73:27	98/96
9	**170a+170b**	Me/Bn/H	51 **(a+b)**	83:17	>99/-

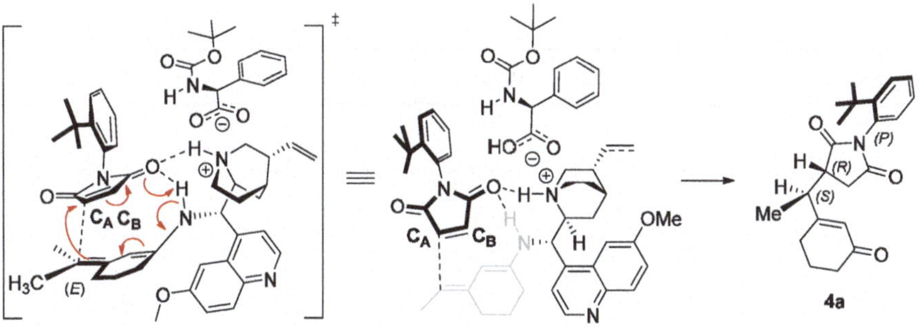

Fig. 3.5 Transition state for the desymmetrization of maleimides with 3-alkylcyclohexenones

ciently promotes the reaction of three classes of nucleophiles, whereas CDA-SQ **172** afforded atropisomeric oxindoles. In all cases, we formed very congested products with adjacent quaternary and tertiary stereocenters and a remote stereogenic axis and in all cases we had complete control over axial chirality. We also investigated the scope of the single nucleophiles starting with 1,3-dicarbonyls (Table 3.4).

The system tolerates both electron withdrawing and donating groups on the aromatic ring of the maleimide (Table 3.4—entries 2–5), sterically demanding *tert*-butyl group does not dramatically influence the efficiency of the desymmetrization (Table 3.4—entries 6–7) and overall high diastereo- and enantioselectivity is

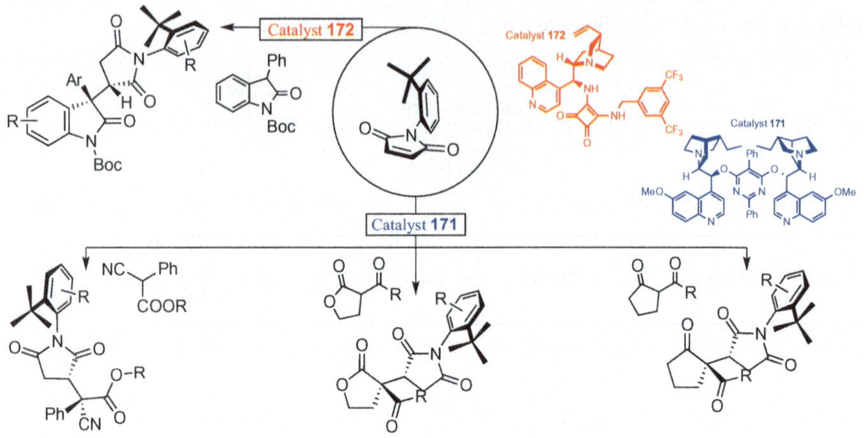

Scheme 3.5 Further development of the desymmetrization strategy

Table 3.4 Scope of α-acyl cyclopentanones

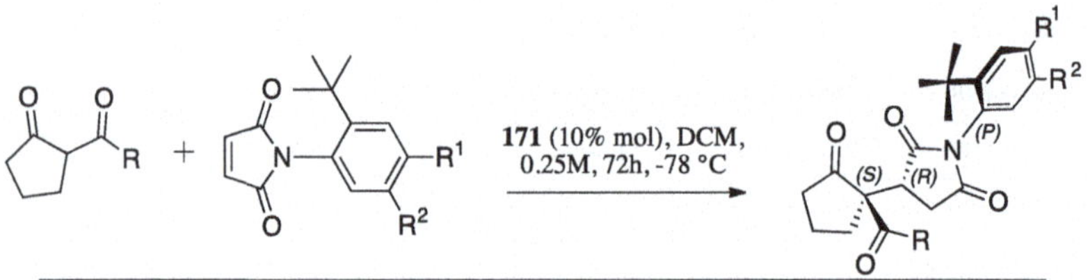

Entry	R	R1	R2	Product	Yield	d.r.	ee
1	Me	H	H	**173**	85	>19:1	93
2	Me	Br	H	**174**	86	17:1	94
3	Me	Cl	H	**175**	81	>19:1	93
4	Me	H	NO_2	**176**	82	>19:1	99
5	Me	OMe	H	**177**	90	19:1	94
6	Me	H	*t*-butyl	**178**	85	18:1	87
7	Me	Br	*t*-butyl	**179**	55	19:1	79
8	Et	H	H	**180**	65	9:1	97
9	Bn	H	H	**181**	36	4:1	37
10	OEt	H	H	**182**	50	8:1	50

achieved. The *P*,*R*,*S*, absolute configuration was assigned to succinimide **179** through single crystal XRD (Fig. 3.6) and the structure obtained showed once more that the nucleophilic attack takes place on the side of the maleimide not shielded by the

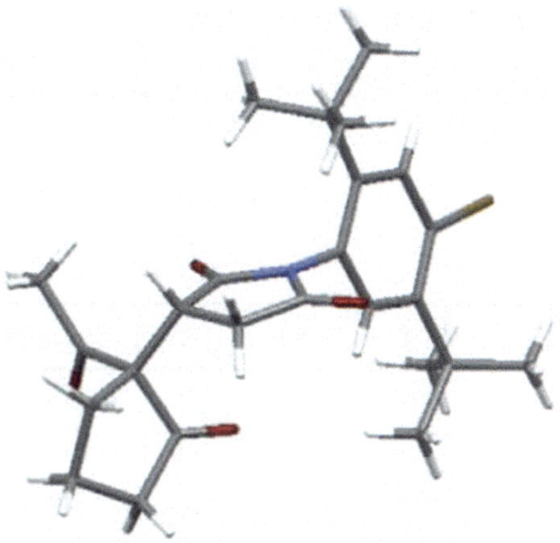

Fig. 3.6 XRD-derived structure of brominated product 179

Table 3.5 Scope of α-acylbutyrolactones

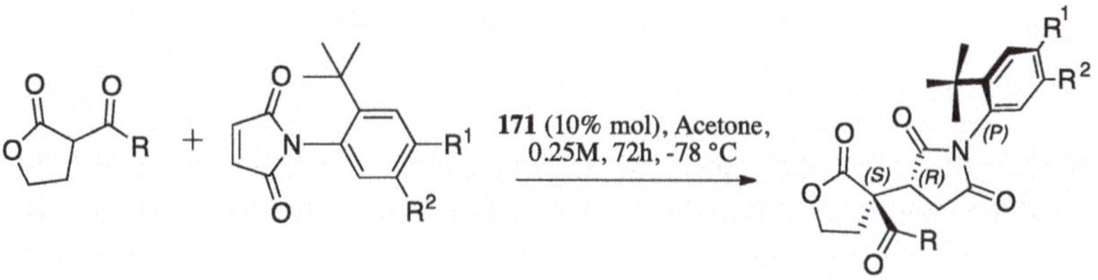

Entry	R	R1	R2	Product	Yield	d.r.	ee
1	Me	H	H	**183**	85	>19:1	96
2	Me	Br	H	**184**	63	19:1	98
3	Me	Cl	H	**185**	93	>19:1	>99
4	Me	H	NO_2	**186**	69	7:1	>99
5	Me	OMe	H	**187**	67	19:1	>99
6	Me	H	*t*-butyl	**188**	71	6:1	90
7	Me	H	NCbz	**189**	90	>19:1	>99
8	4-Br-Ph	H	H	**190**	63	5:1	40
9	4-Br-Ph	H	*t*-butyl	**191**	56	4:1	30

t-butyl group. Other diketones and keto-ester derivatives are reactive under these conditions but with lower efficiency (Table 3.4—entries 8–10).

Next, we reacted some α-acylbutyrolactones with many maleimides under the same type of catalysis but we identified acetone as the best solvent for these substrates (Table 3.5).

The reaction proceeded smoothly for reactivity and selectivity with both electron withdrawing and donating groups (Table 3.5—entries 2–5). In this case the presence

Table 3.6 Scope of the cyanoacetates

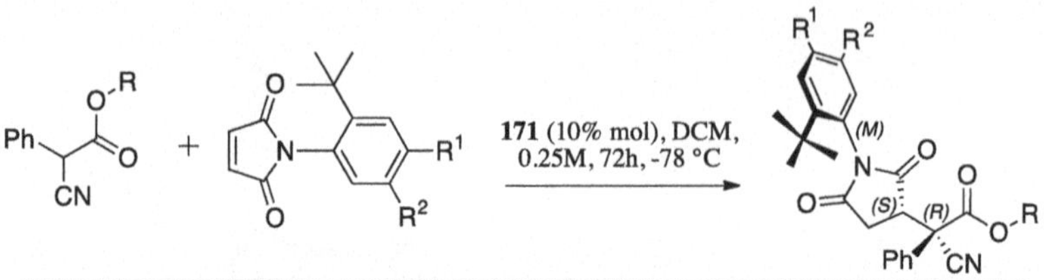

Entry	R	R1	R2	Product	Yield	d.r.	ee
1	*t*-butyl	H	H	**192**	82	>19:1	92
2	*t*-butyl	Br	H	**193**	67	>19:1	81
3	*t*-butyl	Cl	H	**194**	71	>19:1	80
4	*t*-butyl	H	NO_2	**195**	81	13:1	45
5	*t*-butyl	Br	*t*-butyl	**196**	50	19:1	50
6	Et	H	H	**197**	54	19:1	75

of substituents at position 5 of the aromatic ring of the maleimide did not influence the reactivity and the enantioselectivity but had a slight effect on the diastereoselection (Table 3.5—entry 4 and 6). A dramatic drop of efficiency was observed when 2-(4-bromobenzoyl)cyclopentan-1-one was used (Table 3.5—entries 8–9).

Next, we tested the reactivity of cyanoacetates with many maleimides using the same catalysis again in DCM (Table 3.6).

Generally, we found good reactivity and diastereoselectivity, while the enantioselectivity seemed to be strictly dependent by the size of the different substituents of maleimide and phenylacetate.

Finally, we studied the scope of the reaction between 3-aryloxindoles and maleimides. We used milder temperature conditions and bifunctional catalyst **172** providing Brønsted base and H-bonding activations (Table 3.7).

All products were obtained in high enantioselectivity. Specifically, many oxindoles were reacted efficiently (Table 3.7—entries 1–10) and substituents of various nature were tolerated also on the maleimide (Table 3.7—entries 11–16). As a drawback, we observed no reaction when a naphthyl was employed as the aryl group and when 4-substituted oxindoles were used.

3.3 Conclusions

In conclusion, we have successfully achieved the synthesis of axially chiral succinimides via desymmetrization of the corresponding maleimides with many nucleophiles. At the beginning, we explored the feasibility of this strategy using dienamine-

Table 3.7 Scope of 3-aryl oxindoles

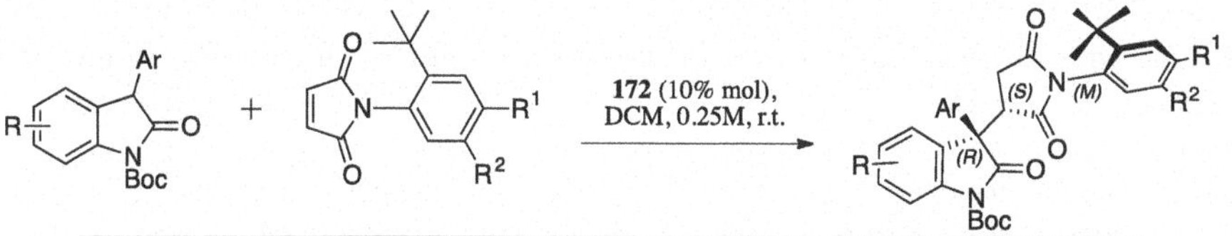

Entry	R	Ar	R1	R2	Product	Yield	d.r.	ee
1	H	Ph	H	H	**198**	82	>19:1	>99
2	5-F	Ph	H	H	**199**	90	10:1	98
3	6-Br	Ph	H	H	**200**	81	19:1	98
4	7-F	Ph	H	H	**201**	98	19:1	98
5	5-OMe	Ph	H	H	**202**	77	>19:1	98
6	5-Me	Ph	H	H	**203**	82	>19:1	98
7	5,7-Me	Ph	H	H	**204**	50	10:1	93
8	H	4-MePh	H	H	**205**	77	10:1	98
9	H	3-MePh	H	H	**206**	79	19:1	98
10	H	4-OMePh	H	H	**207**	82	>19:1	96
11	H	Ph	Cl	H	**208**	90	>19:1	98
12	H	Ph	Br	H	**209**	79	16:1	96
13	H	Ph	Br	*t*-Bu	**210**	50	>19:1	96
14	H	Ph	H	*t*-Bu	**211**	43	>19:1	>99
15	H	Ph	H	NO_2	**212**	81	19:1	98
16	H	Ph	H	NHCbz	**213**	94	6:1	84

activated cyclohexenones as nucleophiles and we were able to completely control the remote axis so we developed this protocol by using other nucleophiles such as 1,3-dicarbonyl or oxindole derivatives activated by a Brønsted base. Thanks to cinchona alkaloid-derived catalysts, we were always able to completely control the axial chirality of the products together with the formation of adjacent quaternary and tertiary stereocenters.

3.4 Experimental Section

3.4.1 General Information

The ^{1}H and ^{13}C NMR spectra were recorded at 400 and 100 MHz, respectively, or at 600 MHz for ^{1}H and 150 MHz for ^{13}C. All the ^{1}H and ^{13}C signals were assigned by means of g-COSY, g-HSQC and g-HMBC 2D-NMR sequences. NOE spectra were recorded using the DPFGSE-NOE sequence, using a mixing time of 1.0–2.0 s and "rsnob" 20–50 Hz wide selective pulses, depending on the crowding of the spectra region. The chemical shifts (δ) for ^{1}H are given in ppm relative to the signals of internal standard TMS and for ^{13}C are given in ppm relative to the signals of the solvents. Coupling constants are given in Hz. When 2D-NMR were not performed, carbon types were determined from DEPT ^{13}C NMR experiments. The following abbreviations are used to indicate the multiplicity: s, singlet; d, doublet; t, triplet; q, quartet; m, multiplet; bs, broad signal. Purification of reaction products was carried out by flash chromatography (FC) on silica gel (230–400 mesh) according to the method of Still [10], or by reverse-phase HPLC (acetonitrile/H_2O mixtures) on C18 columns with a Waters Delta 600 HPLC apparatus equipped with UV detector. Organic solutions were concentrated under reduced pressure on a Büchi rotary evaporator. High Resolution Mass spectra were obtained from the Mass Facility of the Department of Chemistry and Drug Technology of the University of Rome on a Orbitrap Exactive, source: ESI (+): capillary temp: 250 °C, spray voltage: 4.0 (kV), capillary voltage: 65 V, tube lens: 125 V. X-ray data were acquired on a Bruker APEX-2 diffractometer. Optical rotations are reported as follows: $[\alpha]_D^{rt}$ (c in g per 100 mL, solvent). All reactions were carried out in air and using undistilled solvents, without any precautions to exclude moisture unless otherwise noted.

Commercial grade reagents and solvents were used without further purification. Chiral primary amine catalysts **30, 138, 139, 140** and **141** were prepared following the literature procedure [11]. Cyclohexanones were synthesized following the literature procedures [12–16]. Maleimides were prepared following the literature procedures [17, 18]. Oxindoles were prepared following the literature procedures [19, 20].

The diastereomeric ratio was determined by ^{1}H NMR analysis of the crude reaction mixture, and confirmed by HPLC analysis on chiral stationary phases columns. Chiral HPLC analysis was performed on an Agilent 1100-series instrumentation. Phenomenex Lux-Amylose 2 and Phenomenex Lux Cellulose 2 columns Daicel Chiralpak AD-H or AS-H columns and Daicel Chiralcel OD-H with i-PrOH/hexane as the eluent were used. When Chiral HPLC was not successful due to insufficient separation or exceedingly high retention times, enantiomeric excesses were obtained by NMR using enantiopure (*R*)-(-)-2,2,2-trifluoro-1-(9-anthryl)ethanol (Pirkle's alcohol).

3.4.2 *Determination of the Barrier to Racemization of the Chiral Axis for Compound 137a*

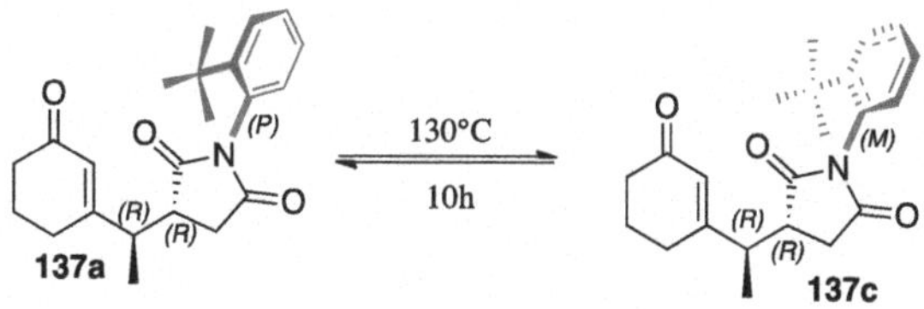

time (s)	x	ln(xe-x)
0	0	0
1800	0,050	-1,1239
3600	0,090	-1,2553
5400	0,135	-1,4271
7200	0,160	-1,5371
9000	0,220	-1,8643
10800	0,230	-1,9310
14400	0,280	-2,3539
18000	0,325	-2,9957
25200	0,350	-3,6889
28800	0,360	-4,1997
∞	0,375	

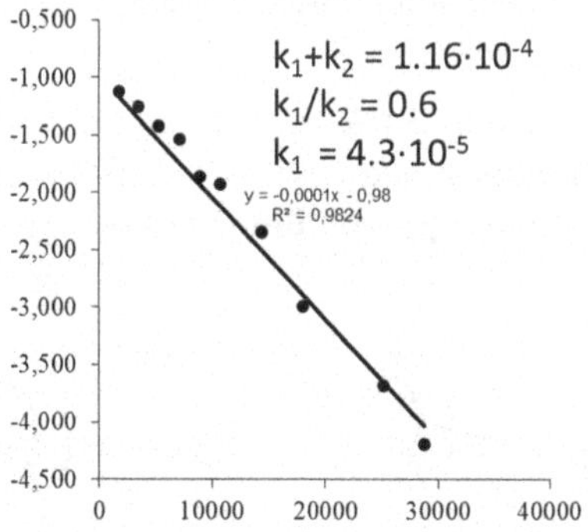

$\Delta G^{\neq} = 31.9$ kcal/mol

Kinetic measurements for the thermal equilibration of the two diastereoisomers (**137a–137c**) arising from the chiral axis of *N*-(*o*-terbutylphenyl). The energy barrier is relative to the **137a** → **137c** conversion. The sample in $C_2D_2Cl_4$ was kept at +130 °C and the NMR measurements of the ratio were obtained at +25 °C.

3.4.3 General Procedure for the Vinylogous Michael Addition of Cyclic Enones to N-Arylmaleimmides

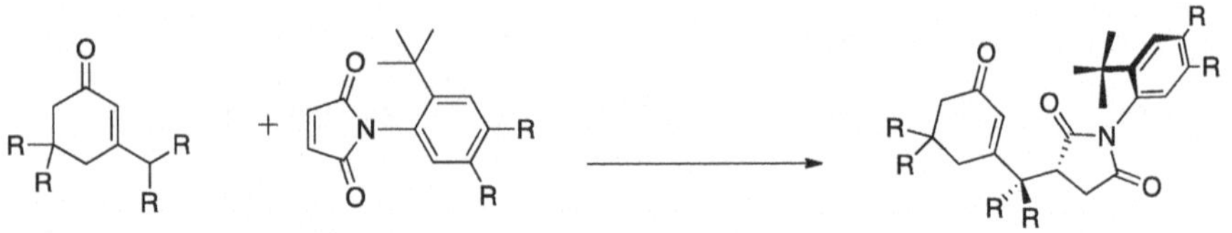

All the reaction were carried out in undistilled toluene. In an ordinary vial equipped with a Teflon-coated stir bar, 9-Amino(9-deoxy)*epi*-quinine **140** (0.04 mmol, 13 mg, 20 mol%) was dissolved in 1.0 mL of toluene and acid cocatalyst **147** (0.08 mmol, 20.1 mg, 40 mol%) was added. The resulting solution was stirred at room temperature for 15 min, then the α,β-unsaturated ketone (0.4 mmol, 2.0 equiv.) was added followed by the maleimide (0.2 mmol, 1 equiv.). The vial was kept stirring at room temperature for 72–96 h. The crude mixture was flushed through a short plug of silica, using dichloromethane/ethyl acetate 1:1 as the eluent (50 ml). Solvent was removed in *vacuo* and the diastereomeric ratio (dr) was determined by ^{1}H NMR analysis of the crude mixture. The desired compound was isolated by flash column chromatography as mixture of two diastereoisomers.

(*P*)-(*R*)-1-(2-(*tert*-butyl)phenyl)-3-((*R*)-1-(3-oxocyclohex-1-en-1-yl)ethyl)pyrrolidine-2,5-dione and **(*P*)-(*R*)-1-(2-(*tert*-butyl)phenyl)-3-((*S*)-1-(3-oxocyclohex-1-en-1-yl)ethyl)pyrrolidine-2,5-dione** (product **148a** and **148b**, Table 3.2—entry 1)

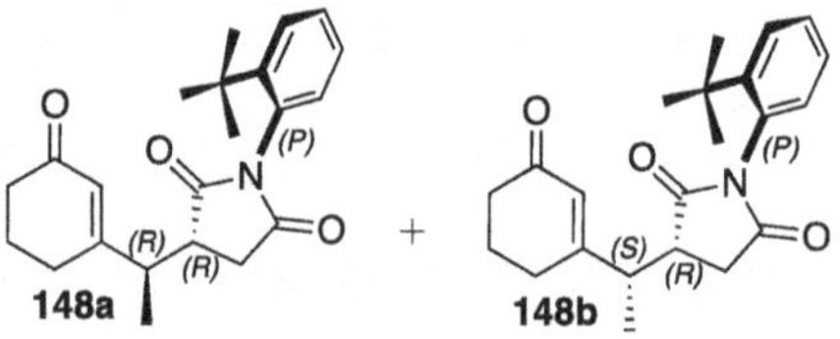

The reaction was carried out at room temperature following the general procedure to furnish the crude product as a 70:30 (^{1}H-NMR signal: δ_{major} 5.92 ppm. bs, δ_{minor} 5.96 ppm. bs) as a mixture of two diastereoisomers **148a** major and **148b** minor. The crude mixture obtained has been purified by flash column chromatography (hexane/ethyl acetate = 3/2) as mixture of **148a** and **148b** in 75% yield and > 99% ee on each diastereoisomer. The ee was determined by HPLC analysis on a Phenomenex Lux-Cellulose 2 column: hexane/*i*-PrOH 80:20, flow rate 1.150 mL/min, λ = 254 nm: **148a** τ_{major} = 36.97 min; **148b** τ_{major} = 34.32 min. The two diastereoisomers have been separated by preparative HPLC on a Kinetex C18 5 μm, 100 Å, 250 × 21.20 mm: acetonitrile/H_2O 50:50, flow rate 20 ml/min, **148a** τ = 9.0 min; **148b** τ = 9.30 min. $[\alpha]_D^{rt}$ on **148a** = −72.0 (*c* = 1.0, $CHCl_3$, > 99% ee). $[\alpha]_D^{rt}$ on **148b** = −37.9 (*c*

= 1.0, $CHCl_3$, > 99% ee). HRMS-ESI-ORBITRAP(+): calculated for $C_{22}H_{27}NO_3$ 376.1883, found 376.1868 $[M+Na]^+$. ^1H NMR of **148a** (600 MHz, $CDCl_3$): δ 1.18 (d, 3H, $J = 6.9$ Hz), 1.30 (s, 9H), 1.99–2.12 (m, 2H), 2.35–2.48 (m, 4H), 2.60 (dd, 1H, $J_1 = 18.7$ Hz, $J_2 = 4.7$ Hz), 2.85 (dd, 1H, $J_1 = 18.7$ Hz, $J_2 = 9.8$ Hz), 3.08–3.15 (m, 1H), 3.27 (m, 1H), 5.92 (bs, 1H), 6.78 (dd, 1H, $J_1 = 7.8$ Hz, $J_2 = 1.5$ Hz), 7.29 (td, 1H, $J_1 = 7.6$ Hz, $J_2 = 1.5$ Hz), 7.41 (m, 1H), 7.59 (dd, 1H, $J_1 = 8.1$ Hz, $J_2 = 1.3$ Hz). ^{13}CNMR (150 MHz, $CDCl_3$): δ 12.9 (CH_3), 22.7 (CH_2), 28.8 (CH_2), 30.6 (CH_2), 31.6 (3 × CH_3), 35.6 (C), 37.5 (CH_2), 40.7 (CH), 42.7 (CH), 126.0 (CH), 127.5 (CH), 129.0 (CH), 130.0 (CH), 130.1 (C), 130.5 (CH), 147.9 (C), 165.6 (C), 176.0 (C), 178.5 (C), 199.3 (C). ^1H NMR of **148b** (600 MHz, $CDCl_3$): (mixture of **148b**:**148a** = 91:9) δ 1.30 (s, 9H), 1.39 (d, 3H, $J = 7.1$ Hz), 1.97–2.10 (m, 2H), 2.32–2.48 (m, 4H), 2.58 (dd, 1H, $J_1 = 18.5$ Hz, $J_2 = 4.9$ Hz), 2.91 (m, 1H), 2.96 (dd, 1H, $J_1 = 18.5$ Hz, $J_2 = 9.6$ Hz), 3.11 (ddd, 1H, $J_1 = 11.9$ Hz, $J_2 = 7.1$, $J_3 = 4.9$ Hz), 5.96 (bs, 1H), 6.76 (dd, 1H, $J_1 = 7.8$ Hz, $J_2 = 1.4$ Hz), 7.29 (m, 1H), 7.40 (m, 1H), 7.59 (dd, 1H, $J_1 = 8.2$ Hz, $J_2 = 1.4$ Hz). ^{13}CNMR (150 MHz, $CDCl_3$): (mixture of **148b**:**148a** = 91:9) δ 17.0 (CH_3), 22.9 (CH_2), 28.4 (CH_2), 31.6 (3 × CH_3), 33.1 (CH_2), 35.7 (C), 37.5 (CH_2), 42.4 (CH), 43.6 (CH), 127.3 (CH), 127.5 (CH), 129.0 (CH), 129.9 (bs, CH+C), 130.4 (CH), 147.9 (C), 165.1 (C), 175.7 (C), 178.1 (C), 199.1 (C).

(*P*)-(*R*)-1-(4-bromo-2-(*tert*-butyl)phenyl)-3-((*R*)-1-(3-oxocyclohex-1-en-1-yl)ethyl)pyrrolidine-2,5-dione and **(*P*)-(*R*)-1-(4-bromo-2-(*tert*-butyl)phenyl)-3-((*S*)-1-(3-oxocyclohex-1-en-1-yl)ethyl)pyrrolidine-2,5-dione** (product **149a** and **149b**, Table 3.2—entry 2)

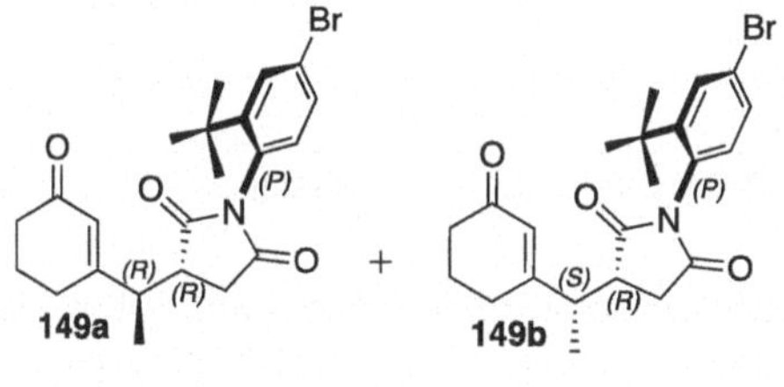

The reaction was carried out at room temperature following the general procedure to furnish the crude product as a 70:30 (^{1}H-NMR signal: δ_{major} 5.88 ppm. bs, δ_{minor} 5.92 ppm. bs) as a crude mixture of two diastereoisomers **149a** major and **149b** minor. The mixture obtained has been purified by flash column chromatography (hexane/ethyl acetate = 3/2) as mixture of **149a** and **149b** in 80% yield and 97% ee on **149a** and > 99% ee on **149b**. The ee was determined by HPLC analysis on a Daicel Chiralcel OD-H column: hexane/*i*-PrOH 90:10, flow rate 1.5 mL/min, $\lambda = 254$ nm: **149a** $\tau_{major} = 36.96$ min, $\tau_{minor} = 27.00$ min; **149b** $\tau_{major} = 49.62$ min. The two diastereoisomers have been separated by preparative HPLC on a Luna C18 5 µm, 100 Å, 250 × 21,20 mm: acetonitrile/H_2O 40:60, flow rate 20 ml/min, **149a** $\tau = 12.30$ min; **149b** $\tau = 15.00$ min. $[\alpha]_D^{rt}$ on **149a** = +51.5 ($c = 1.0$, $CHCl_3$, > 99% ee). $[\alpha]_D^{rt}$ on **149b** = +78.2 ($c = 1.0$, $CHCl_3$, > 99% ee). HRMS-ESI-ORBITRAP (+): calculated for $C_{22}H_{26}BrNO_3$ 432.1169/434.1148, found 432.1163/434.1142 $[M+H]^+$.

^{1}H NMR of **149a** (400 MHz, $CDCl_3$): δ 1.14 (d, 3H, J = 6.9 Hz), 1.27 (s, 9H), 1.93–2.12 (m, 2H), 2.31–2.47 (m, 4H), 2.58 (dd, 1H, J_1 = 18.7 Hz, J_2 = 4.8 Hz), 2.84 (dd, 1H, J_1 = 18.8 Hz, J_2 = 9.8 Hz), 3.08 (m, 1H), 3.25 (m, 1H), 5.89 (bs, 1H), 6.63 (d, 1H, J = 8.3 Hz), 7.40 (dd, 1H, J_1 = 8.2 Hz, J_2 = 2.2 Hz), 7.69 (d, 1H, J = 2.3 Hz). ^{13}C NMR (100 MHz, $CDCl_3$): δ 13.0 (CH_3), 22.7 (CH_2), 28.7 (CH_2), 30.7 (CH_2), 31.4 (3 × CH_3), 35.8 (C), 37.4 (CH_2), 40.7 (CH), 42.7 (CH), 124.2 (C), 126.0 (CH), 129.3 (C), 130.7 (CH), 132.1 (CH), 132.3 (CH), 150.2 (C), 165.4 (C), 175.7 (C), 178.2 (C), 199.2 (C). ^{1}H NMR of **149b** (400 MHz, $CDCl_3$): δ 1.27 (s, 9H), 1.38 (d, 3H, J = 7.0 Hz), 1.93–2.09 (m, 2H), 2.24–2.44 (m, 4H), 2.57 (dd, 1H, J_1 = 18.4 Hz, J_2 = 4.8 Hz), 2.83–2.90 (m, 1H), 2.95 (dd, 1H, J_1 = 18.6 Hz, J_2 = 9.5 Hz), 3.10 (ddd, 1H, J_1 = 12.0 Hz, J_2 = 7.3 Hz, J_3 = 4.9 Hz), 5.94 (bs, 1H), 6.63 (d, 1H, J = 8.4 Hz), 7.41 (dd, 1H, J_1 = 8.4 Hz, J_2 = 2.3 Hz), 2.69 (d, 1H, J = 2.2 Hz). ^{13}CNMR (100 MHz, $CDCl_3$): δ 17.0 (CH_3), 22.8 (CH_2), 28.2 (CH_2), 31.4 (3 × CH_3), 33.2 (CH_2), 35.9 (C), 37.5 (CH_2), 42.7 (CH), 43.5 (CH), 124.2 (C), 127.3 (CH), 129.2 (C), 130.7 (CH), 132.1 (CH), 132.4 (CH), 150.3 (C), 164.9 (C), 175.4 (C), 177.8 (C), 199.0 (C).

(*P*)-(*R*)-1-(2-(*tert*-butyl)-4-chlorophenyl)-3-((*R*)-1-(3-oxocyclohex-1-en-1-yl)ethyl)pyrrolidine-2,5-dione and **(*P*)-(*R*)-1-(2-(*tert*-butyl)-4-chlorophenyl)-3-((*S*)-1-(3-oxocyclohex-1-en-1-yl)ethyl)pyrrolidine-2,5-dione** (product **150a** and **150b**, Table 3.2—entry 3)

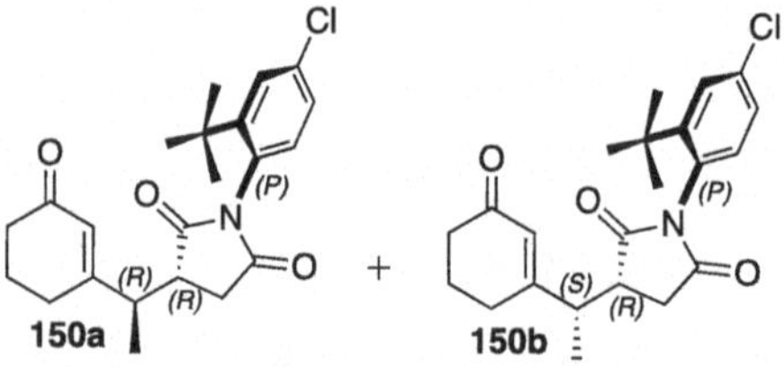

The reaction was carried out at room temperature following the general procedure to furnish the crude product as a 72:28 (^{1}H-NMR signal: δ_{major} 5.91 ppm. bs, δ_{minor} 5.95 ppm. bs) as a mixture of two diastereoisomers **150a** major and **150b** minor. The crude mixture obtained has been purified by flash column chromatography (hexane/ethyl acetate = 7/3) as mixture of **150a** and **150b** isomers in 70% yield and > 99% ee on each diastereoisomer. The ee was determined by HPLC analysis on a Phenomenex Lux Amylose 2 column: hexane/*i*-PrOH 80:20, flow rate 1.5 mL/min, λ = 214, 254 nm: **150a** τ_{major} = 23.93 min; **150b** τ_{major} = 31.85 min. $[\alpha]_D^{rt}$ = +62.7 (c = 1.0, $CHCl_3$, d.r. 68:32, > 99% ee on each isomer). HRMS-ESI-ORBITRAP (+): calculated for $C_{22}H_{26}ClNO_3$ 388.1674, found 388.1670 $[M+H]^+$. ^{1}H NMR (400 MHz, $CDCl_3$): (mixture of **150a**:**150b** = 68:32) δ 1.16 (d, 3.04H, J = 7.1 Hz), 1.27–1.30 (m, 13.23H), 1.39 (d, 3.04H, J = 7.1 Hz), 1.94–2.13 (m, 3.26H), 2.24–2.50 (m, 6.16H), 2.52–2.65 (m, 1.58H), 2.79–3.02 (m, 2.09H), 3.05–3.16 (m, 1.51H), 3.27 (m, 1.02H), 5.91 (bs, 1.0H), 5.95 (bs, 0.47H), 6.69–6.76 (m, 1.48H), 7.22–7.31 (m, 1H), 7.52–7.58 (m, 1.49H). ^{13}CNMR (100 MHz, $CDCl_3$): (mixture of

150a:**150b** = 68:32) δ 12.8 (CH_3), 16.9 (CH_3), 22.7 (CH_2), 22.8 (CH_2), 28.1 (CH_2), 28.6 (CH_2), 30.6 (CH_2), 31.3 (3 × CH_3), 33.1 (CH_2), 35.7 (C), 35.8 (C), 37. 4 (CH_2), 40.6 (CH), 42.4 (CH), 42.7 (CH), 43.4 (CH), 125.9 (CH), 127.2 (CH), 127.6 (CH), 128.5 (C), 128.7 (C), 129.3 (CH), 131.8 (CH), 135.8 (C), 149.9 (C), 165.0 (C), 165.4 (C), 175. 4 (C), 175.7 (C), 177.9 (C), 178.2 (C), 199.1 (C), 199.2 (C).

(*P*)-(*R*)-1-(3-(*tert*-butyl)-[1,1′-biphenyl]-4-yl)-3-((*R*)-1-(3-oxocyclohex-1-en-1-yl)ethyl)pyrrolidine-2,5-dione and **(*P*)-(*R*)-1-(3-(*tert*-butyl)-[1,1′-biphenyl]-4-yl)-3-((*S*)-1-(3-oxocyclohex-1-en-1-yl)ethyl)pyrrolidine-2,5-dione** (product **151a** and **151b**, Table 3.2—entry 4)

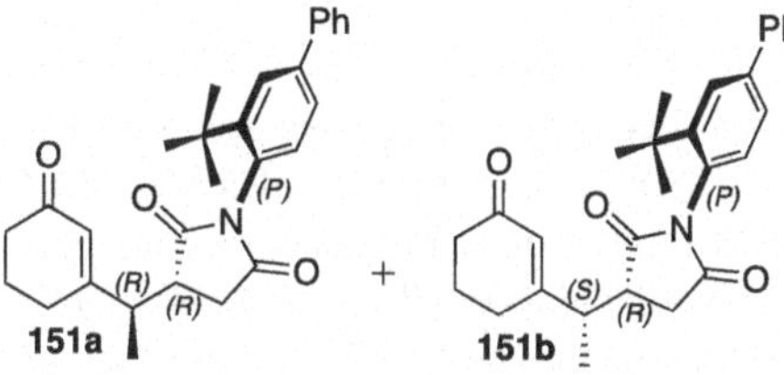

The reaction was carried out at room temperature following the general procedure to furnish the crude product as a 70:30 (^{1}H-NMR signal: δ_{major} 5.93 ppm. bs, δ_{minor} 5.97 ppm. bs) as a mixture of two diastereoisomers **151a** major and **151b** minor. The crude mixture obtained has been purified by flash column chromatography (hexane/acetone = 4/1) as mixture of **151a** and **151b** isomers in 75% yield and 96% ee on **151a** and 95% ee on **151b**. The ee was determined by HPLC analysis on a Daicel Chiralcel OD-H column: hexane/*i*-PrOH 90:10, flow rate 1.2 mL/min, λ = 214, 254 nm: **151a** τ_{major} = 47.7 min, τ_{minor} = 31.9 min; **151b** τ_{major} = 56.2 min, τ_{minor} = 67.7 min. $[\alpha]_D^{rt}$ = +66.1 (c = 1.00, $CHCl_3$, d.r. = 70:30, 96% ee on **151a** and 95% ee on **151b**). HRMS-ESI-ORBITRAP (+): calculated for $C_{28}H_{31}NO_3$ 452.2196, found 452.2194 $[M+H]^+$. ^{1}H NMR (400 MHz, $CDCl_3$): (mixture of **151a**:**151b** = 70:30) δ 1.19 (d, 2.92H, J = 7.0 Hz), 1.33–1.37 (m, 12.02H), 1.40 (d, 1.42H, J = 7.0 Hz), 1.95–2.13 (m, 3.09H), 2.27–2.49 (m, 5.76H), 2.54–2.67 (m, 1.76H), 2.81–3.04 (m, 1.99H), 3.07–3.18 (m, 1.47H), 3.25–3.34 (m, 1.02H), 5.93 (bs, 1H), 5.97 (bs, 0.45H), 6.81–6.88 (m, 1.44H), 7.33–7.41 (m, 1.44H), 7.41–7.51 (m, 4.20H), 7.52–7.59 (m, 2.88H), 7.72–7.80 (m, 1.43H). ^{13}CNMR (100 MHz, $CDCl_3$): (mixture of **151a**:**151b** = 70:30) δ 13.0 (CH_3), 17.0 (CH_3), 22.8 (CH_2), 22.9 (CH_2), 28.3 (CH_2), 28.7 (CH_2), 30.7 (CH_2, 3d), 31.7 (3 × CH_3), 33.2 (CH_2, 4d), 35.8 (C), 35.9 (C), 37.5 (CH_2), 37.6 (CH_2), 40.8 (CH), 42.5 (CH), 42.8 (CH), 43.6 (CH), 126.0 (CH), 126.4 (C+CH), 127.3 (CH), 127.4 (CH), 127.7 (CH), 128.1 (CH), 128.2 (CH), 128.8 (CH), 129.1 (C), 129.2 (C), 130.8 (CH), 130.9 (CH), 140.6 (C), 142.9 (2 × C), 148.1 (C), 148.2 (C), 165.1 (C), 165.6 (C), 175.8 (C), 176.1 (C), 178.3 (C), 178.6 (C), 199.1 (C), 199.3 (C).

(*P*)-(*R*)-1-(2-(*tert*-butyl)-4-methoxyphenyl)-3-((*R*)-1-(3-oxocyclohex-1-en-1-yl)ethyl)pyrrolidine-2,5-dione and **(*P*)-(*R*)-1-(2-(*tert*-butyl)-4-methoxyphenyl)-3-((*S*)-1-(3-oxocyclohex-1-en-1-yl)ethyl)pyrrolidine-2,5-dione** (product **152a** and **152b**, Table 3.2—entry 5)

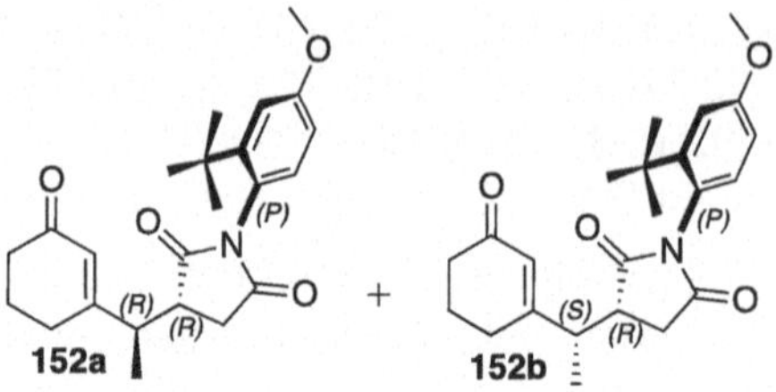

The reaction was carried out at room temperature following the general procedure to furnish the crude product as a 70:30 (^{1}H-NMR signal: δ_{major} 5.89 ppm. bs, δ_{minor} 5.93 ppm. bs) as a mixture of two diastereoisomers **152a** major and **152b** minor. The crude mixture obtained has been purified by flash column chromatography (gradient of hexane/acetone = 4/1 then 7/3) as mixture of **152a** and **152b** isomers in 80% yield and 94% ee on **152a**. The ee of **152a** was determined by ^{1}H-NMR (600 MHz, $CDCl_3$) analysis using 10 equivalents of (*R*)-(-)-2,2,2-Trifluoro-1-(9-anthryl)ethanol at 0 °C: 5.74 (bs, *major enantiomer*) and 5.77 (bs, *minor enantiomer*). $[\alpha]_D^{rt} = +69.5$ (c = 1.0, $CHCl_3$, d.r. 70:30 and 94% ee on **152a**). HRMS-ESI-ORBITRAP (+): calculated for $C_{23}H_{29}NO_4$ 384.2169, found 384.2165 $[M+H]^+$. ^{1}H NMR (400 MHz, $CDCl_3$): (mixture of **152a**:**152b** = 70:30) δ 1.14 (d, 3.15H, J = 6.8 Hz), 1.24–1.29 (m, 14.42H), 1.36 (d, 1.26H, J = 6.8 Hz), 1.92–2.11 (m, 3.11H), 2.24–2.47 (m, 5.88H), 2.50–2.62 (m, 1.60H), 2.74–2.98 (m, 2.01H), 3.02–3.13 (m, 1.49H), 3.23 (m, 1.03H), 7.79 (bs, 4.40H), 5.89 (bs, 1H), 5.93 (bs, 0.46H), 6.66–6.72 (m, 1.45H), 6.77–6.82 (m, 1.53H), 7.06–7.11 (m, 1.46H). ^{13}CNMR (100 MHz, $CDCl_3$): (mixture of **152a**:**152b** = 70:30) δ 13.0 (CH_3), 17.0 (CH_3), 22.7 (CH_2), 22.8 (CH_2), 28.3 (CH_2), 28.7 (CH_2), 30.6 (CH_2), 31.4 (3 × CH_3), 31.5 (3 × CH_3), 33.0 (CH_2), 35.6 (2 × C), 37.4 (CH_2), 37.5 (CH_2), 40.7 (CH), 42.4 (CH), 42.6 (CH), 43.5 (CH), 111.6 (CH), 111.7 (CH), 115.5 (CH), 115.6 (CH), 122.5 (C), 122.7 (C), 125.9 (CH), 127.2 (CH), 131.5 (CH), 131.6 (CH), 149.3 (C), 149.4 (C), 160.2 (2 × C), 165.2 (C), 165.7 (C), 176.1 (C), 176.3 (C), 178.4 (C), 178.8 (C), 199.1 (C), 199.3 (C).

(*P*)-benzyl (4-(*tert*-butyl)-3-((*R*)-2,5-dioxo-3-((*R*)-1-(3-oxocyclohex-1-en-1-yl)ethyl)pyrrolidin-1-yl)phenyl)carbamate and **(*P*)-benzyl (4-(*tert*-butyl)-3-((*R*)-2,5-dioxo-3-((*S*)-1-(3-oxocyclohex-1-en-1-yl)ethyl)pyrrolidin-1-yl)phenyl)carbamate** (product **153a** and **153b**, Table 3.2—entry 6)

@@

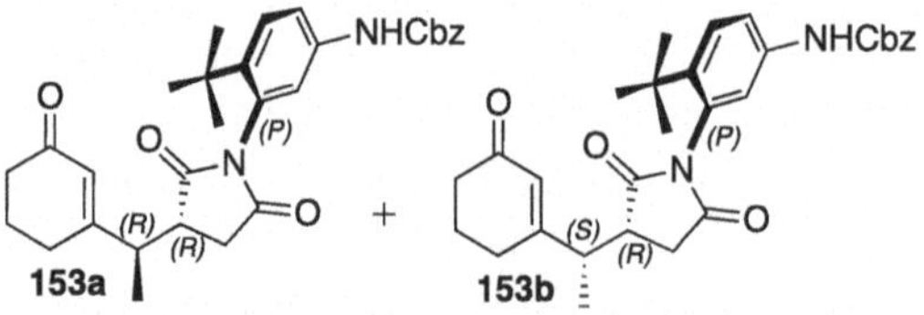

The reaction was carried out at room temperature following the general procedure to furnish the crude product as a 75:25 (^{1}H-NMR signal: δ_{major} 5.89 ppm. bs, δ_{minor}

5.93 ppm. bs) as a mixture of two diastereoisomers **153a** major and **153b** minor. The crude mixture obtained has been purified by flash column chromatography (hexane/ethyl acetate = 7/3) as mixture of **153a** and **153b** isomers in 80% yield and 98% ee on **153a** and 97% ee on **153b**. The ee was determined by HPLC analysis on a Daicel Chiralcel OD-H column: hexane/*i*-PrOH 80:20, flow rate 0.3 mL/min, $\lambda = 214$, 254 nm: **153a** $\tau_{major} = 211.8$ min, $\tau_{minor} = 148.5$ min; **153b** $\tau_{major} = 113.4$ min. $\tau_{minor} = 129.9$ min; $[\alpha]_D^{rt} = +45.0$ ($c = 1.0$, $CHCl_3$, d.r. = 75:25, 98% ee on **153a**, 97% ee on **153b**). HRMS-ESI-ORBITRAP (+): calculated for $C_{30}H_{34}N_2O_5$ 503.2540, found 503.2534 $[M+H]^+$. ^1H NMR (400 MHz, $CDCl_3$): (mixture of **153a**:**153b** = 67/33) δ 1.12 (bd, 2.94H), 1.25–1.27 (m, 15.81H), 1.35 (d, 1.39H, $J = 7.2$ Hz), 1.8–2.09 (m, 4.01H), 2.21–2.47 (m, 5.88H), 2.48–2.65 (m, 1.54H), 2.75–2.97 (m, 2.03H), 2.98–3.12 (m, 1.46H), 3.22 (m, 0.98H), 5.09–5.22 (m, 3.07H), 5.89 (bs, 1H), 5.93 (bs, 0.49H), 6.99–7.16 (m, 3.12H), 7.19–7.27 (m, 1.91H), 7.29–7.38 (m, 7.08H), 7.41–7.47 (m, 1.64H). ^{13}CNMR (100 MHz, $CDCl_3$): (mixture of **153a**:**153b** = 67/33) δ 13.2, 17.1, 22.6, 22.8, 28.4, 28.7, 30.8, 31.5, 31.6, 33.0, 35.2, 37.4, 37.5, 40.8, 42.3, 42.8, 43.6, 67.0, 119.7, 125.9, 127.3, 128.2, 128.3, 128.5, 129.4, 130.3, 130.4, 136.0, 137.2, 142.4, 153.0, 165.3, 165.9, 175.7, 175.8, 178.2, 178.5, 199.4, 199.5.

(*P*)-N-(4-(*tert*-butyl)-3-((*R*)-2,5-dioxo-3-((*R*)-1-(3-oxocyclohex-1-en-1-yl)ethyl)pyrrolidin-1-yl)phenyl)-4-methylbenzenesulfonamide and **(*P*)-N-(4-(*tert*-butyl)-3-((*R*)-2,5-dioxo-3-((*S*)-1-(3-oxocyclohex-1-en-1-yl)ethyl)pyrrolidin-1-yl)phenyl)-4-methylbenzenesulfonamide** (product **154a** and **154b**, Table 3.2—entry 7)

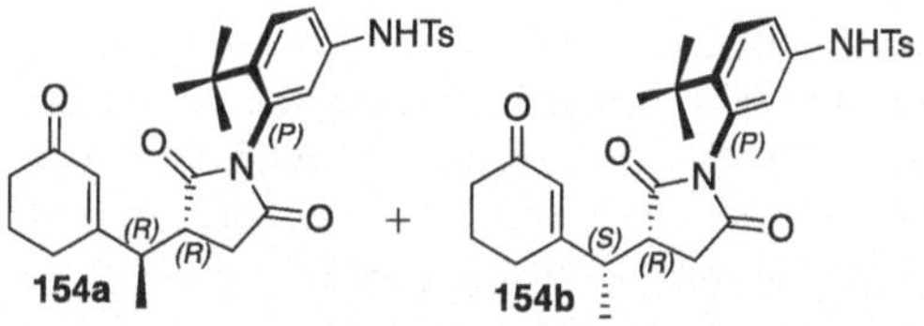

The reaction was carried out at room temperature following the general procedure to furnish the crude product as a 71:29 (^1H-NMR signal: δ_{major} 6.81 ppm. d, δ_{minor} 6.90 ppm. d) as a mixture of two diastereoisomers **154a** major and **154b** minor. The crude mixture obtained has been purified by flash column chromatography (hexane/ethyl acetate = 65/35) as mixture of **154a** and **154b** isomers in 45% yield and 95% ee on each diastereoisomer. The ee of **154a** was determined by ^1H-NMR (600 MHz, $CDCl_3$) analysis using 30 equivalents of (*R*)-(-)-2,2,2-Trifluoro-1-(9-anthryl)ethanol at 0 °C: 5.81 (bs, *major enantiomer*) and 5.78 (bs, *minor enantiomer*) $[\alpha]_D^{rt} = +45.0$ ($c = 1.0$, $CHCl_3$, d.r. = 71:29, 94% ee on **154a**). HRMS-ESI-ORBITRAP (+): calculated for $C_{29}H_{35}N_2O_5S$ 545.2188, found 545.2186 $[M+Na]^+$. ^1H NMR (600 MHz, acetone-d_6): (mixture of **154a**:**154b** = 71:29) δ 1.06 (d, 3.40H, $J = 7.16$ Hz), 1.09 (s, 13.80H), 1.29 (d, 1.35H, $J = 7.00$ Hz), 1.89 (m, 5.20H), 2.20 (m, 3.21H), 2.28–2.34 (m, 2.12H), 2.38–2.43 (m, 1.18H), 2.52 (m, 1.64H), 2.70 (s, 2.88H), 2.92 (m, 3.14H), 3.22 (m, 0.46H), 3.39 (m, 1.07H), 5.76 (bs, 1.45H), 6.65 (d, 1.00H, $J = 2.63$ Hz), 6.74 (d, 0.44H, $J = 2.44$ Hz), 7.01 (dd, 0.49H, $J_1 = 8.76$ Hz, $J_2 = 2.57$ Hz), 7.07 (dd, 1.12H, $J_1 = 8.58$ Hz, $J_2 = 2.39$ Hz), 7.20 (bs, 3.20H),

7.33 (dd, 1.58H, $J_1 = J_2 = 9.21$ Hz), 7.55 (bs, 2.98H), 8.83 (s, 0.39H), 8.91 (s, 0.91H). ^{13}CNMR (150 MHz, acetone-d$_6$): (mixture of **154a**:**154b** = 71:29) δ 14.1, 14.8, 17.8, 21.9, 23.8, 24.1, 24.2, 28.7, 29.4, 32.1, 32.3, 32.8, 34.8, 36.2, 36.3, 38.7, 38.7, 42.2, 43.8, 44.1, 44.5, 122.0, 122.3, 123.5, 123.9, 126.8, 127.9, 128.5, 130.5, 130.6, 131.0, 131.0, 133.3, 133.4, 138.1, 138.2, 138.4, 138.5, 145.1, 145.1, 145.2, 145.3, 167.3, 167.7, 177.0, 177.2, 179.8, 179.9, 199.2, 199.3.

(*P*)-(*R*)-1-(2-(*tert*-butyl)phenyl)-3-((*R*)-1-(5,5-dimethyl-3-oxocyclohex-1-en-1-yl)ethyl)pyrrolidine-2,5-dione and **(*P*)-(*R*)-1-(2-(*tert*-butyl)phenyl)-3-((*S*)-1-(5,5-dimethyl-3-oxocyclohex-1-en-1-yl)ethyl)pyrrolidine-2,5-dione** (product **155a** and **155b**, Table 3.2—entry 8)

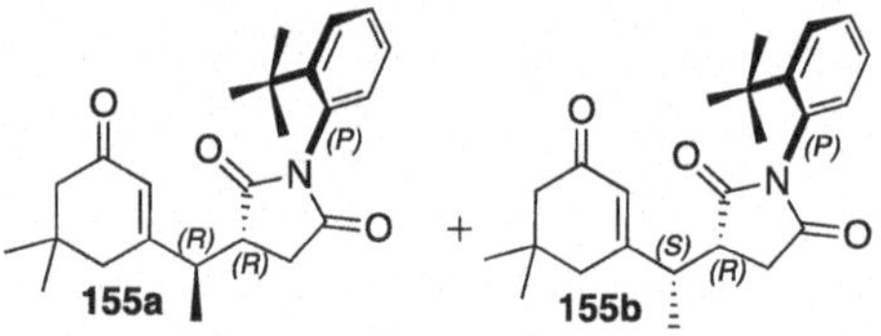

The reaction was carried out at room temperature following the general procedure t to furnish the crude product as a 70:30 (^{1}H-NMR signal: δ_{major} 5.93 ppm. bs, δ_{minor} 5.97 ppm. bs) as a mixture of two diastereoisomers **155a** major and **155b** minor. The crude mixture obtained has been purified by flash column chromatography (hexane/ethyl acetate = 70/30) as mixture of **155a** and **155b** isomers in 60% yield and 96% ee on **155a** and 95% ee on **155b**. The ee was determined by HPLC analysis on a Daicel Chiralpak AD-H column: hexane/*i*-PrOH 95:5, flow rate 0.75 mL/min, $\lambda = 214, 254$ nm: **155a** $\tau_{major} = 55.6$ min, $\tau_{minor} = 58.9$ min; **155b** $\tau_{major} = 62.1$ min, $\tau_{minor} = 54.2$ min. $[\alpha]_D^{rt} = +69.0$ (c = 1.0, $CHCl_3$, d.r. 70:30, 96% ee on **155a** and 95% ee on **155b**). HRMS-ESI-ORBITRAP (+): calculated for $C_{24}H_{31}NO_3$ 382.2377, found 523.2368 $[M+H]^+$. ^{1}H NMR (600 MHz, $CDCl_3$): (mixture of **155a**:**155b** = 70/30) δ 1.05 (s, 4.60H), 1.09 (s, 4.79H), 1.17 (d, 3.40H, J = 6.92 Hz), 1.31 (bs, 14.68H), 1.40 (d, 1.94H, J = 7.55 Hz), 2.28 (bs, 6.30H), 2.59 (m, 1.74H), 2.82 (m, 1.74H), 2.97 (m, 0.60H), 3.10 (m, 1.64H), 3.27 (m, 1.10H), 5.93 (s, 1.00H), 5.97 (s, 0.46H), 6.78 (m, 1.35H), 7.29 (m, 1.55H), 7.41 (m, 1.55H), 7.59 (m, 1.54H). ^{13}CNMR (150 MHz, $CDCl_3$): (mixture of **155a**:**155b** = 70/30) δ 12.5, 14.1, 16.9, 22.6, 27.6, 27.8, 28.4, 28.7, 30.4, 31.5, 31.6, 33.4, 33.7, 35.6, 40.3, 42.1, 42.4, 42.7, 42.9, 43.3, 51.0, 51.1, 124.9, 126.1, 127.4, 127.5, 128.9, 129.0, 129.9, 129.9, 130.1, 130.4, 130.5, 147.9, 162.7, 163.1, 175.7, 178.1, 178.5, 199.4.

(*P*)-(*R*)-1-(2-(*tert*-butyl)-4-chlorophenyl)-3-((*R*)-1-(5,5-dimethyl-3-oxocyclohex-1-en-1-yl)ethyl)pyrrolidine-2,5-dione and **(*P*)-(*R*)-1-(2-(*tert*-butyl)-4-chlorophenyl)-3-((*S*)-1-(5,5-dimethyl-3-oxocyclohex-1-en-1-yl)ethyl)pyrrolidine-2,5-dione** (product **156a** and **156b**, Table 3.2—entry 9)

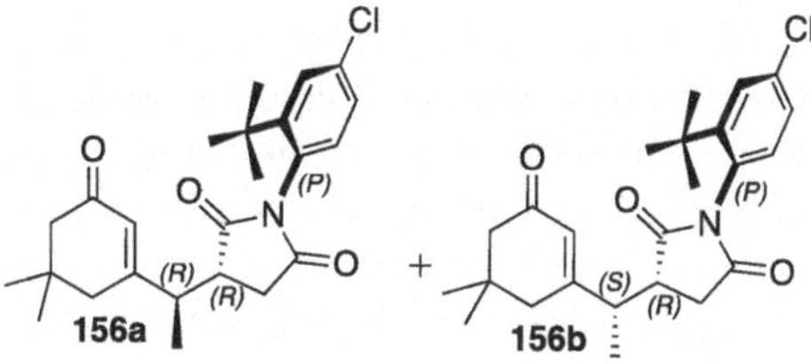

The reaction was carried out at room temperature following the general procedure to furnish the crude product as a 65:35 (^{1}H-NMR signal: δ_{major} 5.92 ppm. bs, δ_{minor} 5.95 ppm. bs) as a mixture of two diastereoisomers **156a** major and **156a** minor. The crude mixture obtained has been purified by flash column chromatography (hexane/ethyl acetate = 7/3) as mixture of **156a** and **156b** isomers in 60% yield and 95% ee on each diastereoisomer. The ee was determined by HPLC analysis on a Daicel Chiralpak AD-H column: hexane/*i*-PrOH 90:10, flow rate 0.75 mL/min, λ = 214, 254 nm: **156a** τ_{major} = 11.79 min, τ_{minor} = 9.61 min; **156b** τ_{major} = 16.58 min, τ_{minor} = 19.70 min. $[\alpha]_D^{rt}$ = +58.2 (*c* = 1.0, $CHCl_3$, d.r. 70:30, 95% ee on **156a** and 95% ee on **156b**). HRMS-ESI-ORBITRAP (+): calculated for $C_{24}H_{30}ClNO_3$ 416.1987, found 416.1981 $[M+H]^+$. ^{1}H NMR (400 MHz, $CDCl_3$): (mixture of **156a**:**156b** = 69:31) δ 1.01–1.11 (m, 8.63H), 1.15 (d, 3.18H, *J* = 6.9 Hz), 1.27–1.31 (m, 13.19H), 1.39 d, 1.32H, *J* = 6.9 Hz), 2.14–2.35 (m, 6.40H), 2.51–2.65 (m, 1.68H), 2.74–2.89 (m, 1.51H), 2.89–3.03 (m, 0.54H), 3.03–3.15 (m, 1.47H), 3.20–3.33 (m, 1.0H), 5.92 (bs, 1H), 5.95 (bs, 0.45H), 6.67–6.77 (m, 1.48H), 7.23–7.31 (m, 1.81H), 7.51–7.58 (m, 1.46H). ^{13}CNMR (100 MHz, $CDCl_3$): (mixture of **156a**:**156b** = 69:31) δ 12.5 (CH_3), 16.9 (CH_3), 27.7 (CH_2), 27.8 (CH_2), 28.4 (CH_2), 28.7 (CH_2), 30.4 (CH_2), 31.4 (3 × CH_3) 33.4 (CH_2), 33.7 (C), 35.8 (C), 35.9 (C), 40.3 (CH), 42.1 (CH_2), 42.4 (CH), 42.7 (CH), 43.0 (CH_2), 43.3 (CH), 125.0 (CH) 126.2 (CH), 127.6 (CH), 127.7 (CH), 128.6 (C), 128.8 (C), 129.3 (CH), 129.4 (CH), 131.8 (CH), 131.9 (CH), 135.9 (2 × C), 150.0 (2 × C), 162.5 (C), 162.9 (C), 175.4 (C), 175.7 (C), 177.9 (C), 178.4 (C), 199.3 (C), 199.4 (C).

(*P*)-(*R*)-1-(2-(*tert*-butyl)phenyl)-3-((*R*)-1-(3-oxocyclohex-1-en-1-yl)-2-phenylethyl)pyrrolidine-2,5-dione and **(*P*)-(*R*)-1-(2-(*tert*-butyl)phenyl)-3-((*S*)-1-(3-oxocyclohex-1-en-1-yl)-2-phenylethyl)pyrrolidine-2,5-dione** (product **157a** and **157b**, Table 3.2—entry 10)

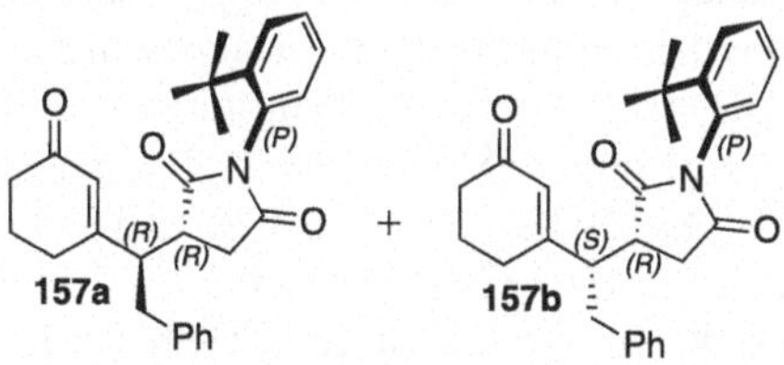

The reaction was carried out at room temperature following the general procedure to furnish the crude product as a 70:30 (^{1}H-NMR signal: δ_{major} 6.76 ppm. bs, δ_{minor} 6.63 ppm. d as a mixture of two diastereoisomers **157a** major and **157b** minor. The crude mixture obtained has been purified by flash column chromatography (hexane/ethyl acetate = 7/3) as mixture of **157a** and **157b** isomers in 50%

yield and 99% ee on **157a** and 97% ee on **157b**. The ee was determined by HPLC analysis on a Phenomenex Lux-Cellulose 2 column: hexane/*i*-PrOH 80:20, flow rate 1.0 mL/min, $\lambda = 214$ nm: **157a** $\tau_{major} = 39.2$ min; $\tau_{minor} = 34.7$ min; **157b** $\tau_{major} = 45.6$ min; $\tau_{minor} = 53.4$ min. $[\alpha]_D^{rt} = +44.9$ ($c = 0.94$, $CHCl_3$, d.r. 70:30, 99% ee on **157a** and 97% ee on **157b**). HRMS-ESI-ORBITRAP (+): calculated for $C_{28}H_{31}NO_3$ 430.2377, found 430.2376 $[M+H]^+$. ^{1}H NMR (600 MHz, $CDCl_3$): (mixture of **157a**:**157b** = 67:33) δ 1.28 (bs, 15,64H), 1.91 (m, 3.18H), 2.15 (m, 1.67H), 2.35 (4.75H), 2,68 (dd, 1.1H, $J_1 = 19.00$ Hz, $J_2 = 4.46$ Hz), 2.80 (dd, 0.60H, $J_1 = 18.36$ Hz, $J_2 = 5.41$ Hz), 2.96 (m, 3.65H), 3.14 (m, 2.57H), 3.23 (m, 0.50H), 3.29 (dd, 0.98H, $J_1 = 13.48$ Hz, $J_2 = 5.26$ Hz), 5.90 (s, 1.01H), 5.93 (s, 0.47H), 6.63 (d, 0.49H, $J = 7.48$ Hz), 6.76 (d, 1.00H, $J = 7.72$ Hz), 7.17 (m, 3.23H), 7.24 (m, 3.22H), 7.29 (m, 4.17H), 7.40 (m, 1.53H), 7.58 (m, 1.46H). ^{13}CNMR (150 MHz, $CDCl_3$): (mixture of **157a**:**157b** = 67:33) δ 22.5, 22.6, 29.4, 29.7, 30.0, 31.6, 32.4, 32.5, 35.6, 35.6, 36.2, 37.3, 37.4, 37.9, 42.2, 43.0, 48.9, 49.3, 126.9, 127.4, 127.5, 127.7, 128.7, 128.7, 128.8, 128.9, 128.9, 129.0, 129.9, 129.9, 130.4, 130.5, 137.6, 137.9, 147.8, 147.9, 163.5, 164.2, 175.4, 175.6, 177.9, 178.4, 198.6, 199.0.

(*P*)-(*R*)-1-(2-(*tert*-butyl)phenyl)-3-((*R*)-1-(3-oxocyclohex-1-en-1-yl)butyl)pyrrolidine-2,5-dione and **(*P*)-(*R*)-1-(2-(*tert*-butyl)phenyl)-3-((*S*)-1-(3-oxocyclohex-1-en-1-yl)butyl)pyrrolidine-2,5-dione** (product **158a** and **158b**, Table 3.2—entry 11)

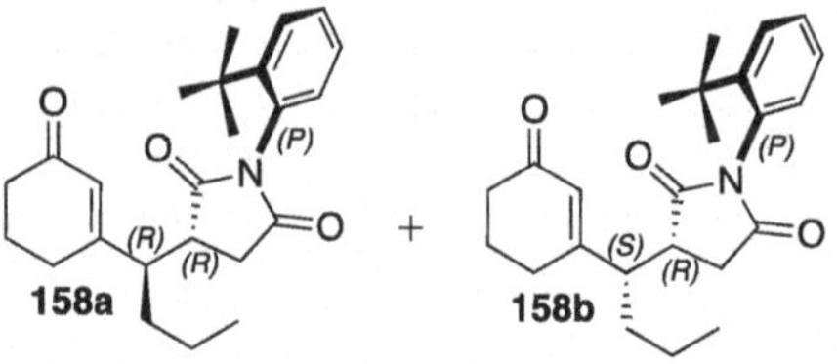

The reaction was carried out at room temperature following the general procedure to furnish the crude product as a 64:36 (^{1}H-NMR signal: δ_{major} 6.73 ppm. bs, δ_{minor} 6.77 ppm. bs) as a mixture of two diastereoisomers **158a** major and **158b** minor. The crude mixture obtained has been purified by flash column chromatography (hexane/ethyl acetate = 7/3) as mixture of **158a** and **158b** isomers in 37% yield and 97% ee on **158a** and 95% ee on **158b**. The ee was determined by HPLC analysis on a Daicel Chiralpak AD-H column: hexane/*i*-PrOH 90:10, flow rate 0.75 mL/min, $\lambda = 214$, 254 nm: **158a** $\tau_{major} = 27.4$ min, $\tau_{minor} = 18.1$ min; **158b** $\tau_{major} = 21.7$ min, $\tau_{minor} = 19.5$ min2. $[\alpha]_D^{rt} = +40.8$ ($c = 1.0$, $CHCl_3$, d.r. 70:30, 97% ee on **158a** and 95% ee on **158b**). HRMS-ESI-ORBITRAP (+): calculated for $C_{24}H_{31}NO_3$ 382.2377, found 382.2371 $[M+H]^+$. ^{1}H NMR (400 MHz, $CDCl_3$): (mixture of **158a**:**158b** = 67:33) δ 0.94 (m, 4.83H), 1.30 (m, 20.50H), 1.64 (m, 2.36H), 2.00 (m, 3.98H), 2.10 (m, 0.64H), 2.27 (m, 1.24H), 2.40 (m, 5.47H), 2.63 (dd, 1.14H, $J_1 = 18.54$ Hz, $J_2 = 5.06$ Hz), 2.70 (dd, 0.57H, $J_1 = 18.31$ Hz, $J_2 = 4.99$ Hz), 2.79 (m, 1.61H), 2.92 (m, 1.62H), 3.12 (m, 1.59H), 5.94 (s, 0.49H), 5.96 (s, 1.01H), 6.73 (d, 1.00H, $J = 7.74$ Hz), 6.77 (d, 0.49H, $J = 7.74$ Hz), 7.29 (m, 2.27H), 7.40 (m, 1.57H), 7.58 (d,

1.53H, $J = 8.14$ Hz). ^{13}CNMR (100 MHz, $CDCl_3$): (mixture of **158a**:**158b** = 67:33) δ 13.9, 14.0, 20.5, 20.6, 22.7, 22.8, 28.4, 28.5, 29.7, 31.4, 31.6, 32.3, 32.9, 33.1, 35.6, 35.7, 37.5, 37.6, 42.9, 43.6, 47.5, 47.9, 127.4, 2 × 127.5, 128.8, 2 × 129.0, 3129.9, 130.0, 130.4, 130.5, 147.9, 147.9, 163.9, 164.8, 175.7, 175.9, 178.0, 178.4, 198.9, 199.3.

(*P*)-(*R*)-1-(2-(*tert*-butyl)phenyl)-3-((*R*)-1-(3-oxocyclohex-1-en-1-yl)but-3-en-1-yl)pyrrolidine-2,5-dione and **(*P*)-(*R*)-1-(2-(*tert*-butyl)phenyl)-3-((*S*)-1-(3-oxocyclohex-1-en-1-yl)but-3-en-1-yl)pyrrolidine-2,5-dione** (product **159a** and **159b**, Table 3.2—entry 12)

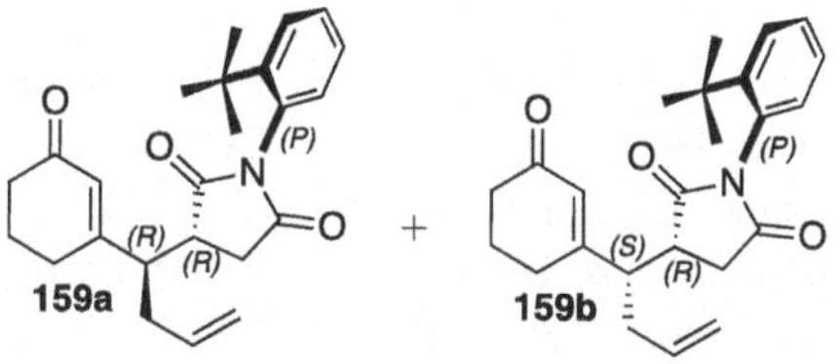

The reaction was carried out at room temperature following the general procedure to furnish the crude product as a 55:45 (^{1}H-NMR signal: δ_{major} 6.73 ppm. d, δ_{minor} 6.79 ppm. d) as a mixture of two diastereoisomers **159a** major and **159b** minor. The crude mixture obtained has been purified by flash column chromatography (hexane/ethyl acetate = 7/3) as mixture of **159a** and **159b** isomers in 36% yield and 96% ee on each diastereoisomer. The ee was determined by HPLC analysis on a Daicel Chiralpak AD-H column: hexane/*i*-PrOH 90:10, flow rate 0.75 mL/min, λ = 214, 254 nm: **159a** τ_{major} = 28.7 min, τ_{minor} = 20.4 min; **159b** τ_{major} = 31.5 min, τ_{minor} = 27.0 min. $[\alpha]_D^{rt}$ = +48.0 (c = 1.0, $CHCl_3$, d.r. = 70:30, 96% ee on each isomer). HRMS-ESI-ORBITRAP (+): calculated for $C_{24}H_{29}NO_3$ 380.2220, found 380.2219 $[M+H]^+$. ^{1}H NMR (400 MHz, $CDCl_3$): (mixture of **159a**:**159b** = 55:45) δ 1.29 (bs, 18.68 Hz), 2.04 (m, 4.34H), 2.39 (m, 11.68H), 2.62 (dd, 1.13H, J_1 = 18.69 Hz, J_2 = 5.41 Hz), 2.68 (m, 1.15H), 2.75 (dd, 1.01H, J_1 = 18.59 Hz, J_2 = 5.41 Hz), 2.94 (m, 4.14H), 3.17 (m, 2.02H), 5.14 (m, 4.01H), 5.70 (m, 1.93H), 5.49 (bs, 1.91H), 6.73 (d, 1H, J = 7.63 Hz), 6.79 (d, 0.90H, J = 7.63 Hz), 7.29 (m, 2.12H), 7.40 (m, 1.99H), 7.58 (m, 1.95H). ^{13}CNMR (100 MHz, $CDCl_3$): (mixture of **159a**:**159b** = 55:45) δ 22.6, 22.7, 28.8, 29.06, 29.7, 31.6, 31.8, 32.5, 34.1, 35.5, 35.6, 35.7, 37.5, 37. 6, 42.1, 42.9, 46.8, 47.1, 118.2, 118.3, 127.4, 2 × 127.5, 128.7, 2 × 129.0, 2 × 129.9, 130.0, 130.3, 130.4, 134.4, 147.8, 147.9, 163.2, 164.4, 175.5, 175.7, 178.00, 178.4, 198.8, 199.2.

(*P*)-(*R*)-1-(4-bromo-2-(*tert*-butyl)phenyl)-3-((*R*)-1-(3-oxocyclohex-1-en-1-yl)butyl)pyrrolidine-2,5-dione and **(*P*)-(*R*)-1-(4-bromo-2-(*tert*-butyl)phenyl)-3-((*S*)-1-(3-oxocyclohex-1-en-1-yl)butyl)pyrrolidine-2,5-dione** (product **160a** and **160b**, Table 3.2—entry 13)

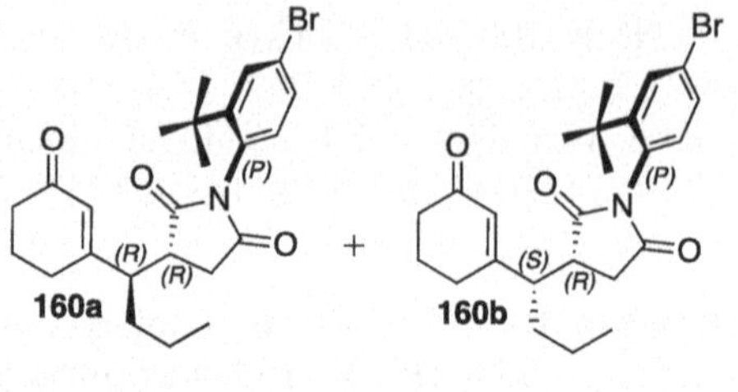

The reaction was carried out at room temperature following the general procedure to furnish the crude product as a 70:30 (^{1}H-NMR signal: δ_{major} 5.95 ppm. bs, δ_{minor} 5.92 ppm. bs) as a mixture of two diastereoisomers **160a** major and **160b** minor. The crude mixture obtained has been purified by flash column chromatography (hexane/ethyl acetate = 7/3) as mixture of **160a** and **160b** isomers in 76% yield and 99% ee on **160a** and 98% ee on **160b**. The ee was determined by HPLC analysis on a Daicel Chiralpak AS-H column: hexane/*i*-PrOH 90:10, flow rate 1.2 mL/min, $\lambda = 214, 254$ nm: **160a** $\tau_{major} = 31.5$ min, $\tau_{minor} = 37.2$ min; **160b** $\tau_{major} = 63.9$ min, $\tau_{minor} = 46.0$ min. $[\alpha]_D^{rt} = +33.4$ ($c = 1.0$, $CHCl_3$, d.r. = 70:30, 99% ee on **160a** and 98% ee on **160b**). HRMS-ESI-ORBITRAP (+): calculated for $C_{24}H_{30}BrNO_3$ 460.1482/462.1462, found 460.1476/460.1455 $[M+H]^+$. ^{1}H NMR (600 MHz, $CDCl_3$): (mixture of **160a**:**160b** = 70:30) δ 0.88 (t, 2.02H, $J = 7.17$ Hz), 0.94 (m, 4.50H), 1.28 (bs, 20.4H), 1.64 (m, 2.63H), 1.99 (m, 3.76H), 2.10 (m, 0.58H), 2.26 (m, 1.18H), 2.36 (m, 2.18H), 2.42 (m, 2.98H), 2.62 (dd, 1.11H, $J_1 = 18.55$ Hz, $J_2 = 5.11$ Hz), 2.70 (dd, 0.51H, $J_1 = 18.55$ Hz, $J_2 = 5.11$ Hz), 2.74 (m, 1.54H), 2.92 (m, 1.54H), 3.11 (m, 1.53H), 5.92 (s, 0.45H), 5.95 (s, 1.00H), 6.61 (d, 0.98H, $J = 8.28$ Hz), 6.64 (d, 0.45H, $J = 8.28$ Hz), 7.42 (m, 1.42H), 7.70 (d, 1.41H, $J = 2.08$ Hz). ^{13}CNMR (150 MHz, $CDCl_3$): (mixture of **160a**:**160b** = 70:30) δ 12.9, 13.0, 13.1, 19.4, 19. 6, 21.6, 21.7, 21.8, 27.3, 27.5, 2 × 30.4, 30.6, 31.4, 31.9, 32.1, 2 × 34.8, 36.5, 36.6, 41.9, 42.6, 46.5, 46.9, 123.2, 123.3, 126.4, 127.8, 128.1, 128.2, 2 × 129.70, 131.1, 2 × 131.3, 149.2, 149.3, 162.8, 163.6, 174.4, 174.6, 176.7, 177.1, 197.8, 198.2.

(*P*)-benzyl (4-(*tert*-butyl)-3-((*R*)-2,5-dioxo-3-((*R*)-1-(3-oxocyclohex-1-en-1-yl)but-3-en-1-yl)pyrrolidin-1-yl)phenyl)carbamate and **(*P*)-benzyl (4-(*tert*-butyl)-3-((*R*)-2,5-dioxo-3-((*S*)-1-(3-oxocyclohex-1-en-1-yl)but-3-en-1-yl)pyrrolidin-1-yl)phenyl)carbamate** (product **161a** and **161b**, Table 3.2—entry 14)

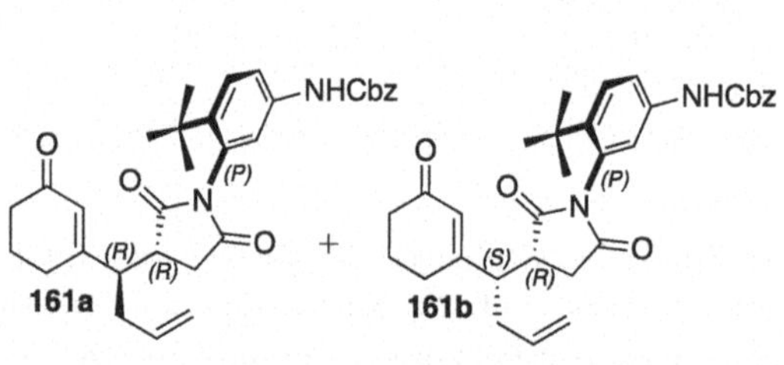

The reaction was carried out at room temperature following the general procedure to furnish the crude product as a 70:30 (^{1}H-NMR signal: δ_{major} 1.24 ppm. bs, δ_{minor} 1.25 ppm. bs) as a mixture of two diastereoisomers **161a** major and **161b**

minor. The crude mixture obtained has been purified by flash column chromatography (hexane/ethyl acetate = 3/2) as mixture of **161a** and **161b** isomers in 63% yield and 97% ee on each diastereoisomer. The ee was determined by HPLC analysis on a Daicel Chiralpak AD-H column: hexane/*i*-PrOH 80:20, flow rate 1.0 mL/min, $\lambda = 214, 254$ nm: **161a** $\tau_{major} = 47.8$ min, $\tau_{minor} = 26.2$ min; **161b** $\tau_{major} = 29.5$ min, $\tau_{minor} = 15.8$ min. $[\alpha]_D^{rt} = +18.0$ (c = 1.0, $CHCl_3$, d.r. = 70:30, 97% ee on each isomer). HRMS-ESI-ORBITRAP (+): calculated for $C_{32}H_{36}N_2O_5$ 529.2697, found 529.2691 $[M+H]^+$. ^1H NMR (400 MHz, $CDCl_3$): (mixture of **161a**:**161b** = 70:30) δ 1.18–1.32 (m, 9H), 1.89–2.12 (m, 2H), 2.16–2.50 (m, 5.5H), 2.51–2.77 (m, 1.5H), 2.80–3.02 (m, 2H), 3.07–3.20 (m, 1H), 5.04–5.23 (m, 4H), 5.59–5.79 (m, 1H), 5.92 (bs, 1H), 6.94–7.04 (m, 1H), 7.04–7.16 (m, 1H), 7.19–7.28 (m, 1H), 7.29–7.40 (m, 5H), 7.40–7.48 (m, 1H). ^{13}CNMR (100 MHz, $CDCl_3$): (mixture of **161a**:**161b** = 70:30) δ 22.6, 22.7, 29.0, 29.2, 31.6, 31.9, 32.3, 34.5, 35.2, 35.3, 35.5, 37.5, 37.6, 42.1, 42.8, 53.4, 67.0, 118.2, 118.6, 119.6, 127.3, 128.2, 128.3, 128.6, 128.7, 129.3, 129.4, 130.1, 130.3, 130.4, 134.4, 134.5, 2 × 136.0, 2 × 137.2, 153.0, 163.4, 164.8, 175.5, 175.6, 178.0, 178.4, 199.3, 199.5.

(*P*)-(*R*)-1-(2-(tert-butyl)phenyl)-3-(2-(3-oxocyclohex-1-en-1-yl)propan-2-yl)pyrrolidine-2,5-dione (product **162**, Table 3.3—entry 1)

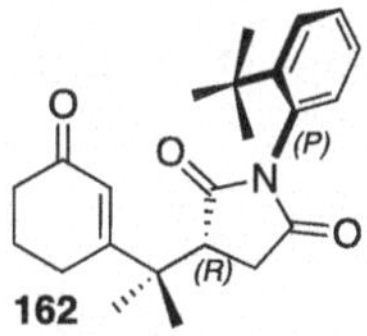

162

The reaction was carried out at room temperature following the general procedure furnish the crude product **162**. The crude mixture obtained has been purified by flash column chromatography (hexane/ethyl acetate = 3/2) in 83% yield and 96% ee. The ee was determined by HPLC analysis on a Daicel Chiralpak AD-H column: hexane/*i*-PrOH 90:10, flow rate 1.0 mL/min, $\lambda = 214, 254$ nm: **162** $\tau_{major} = 18.1$ min, $\tau_{minor} = 21.3$ min. $[\alpha]_D^{rt} = +33.0$ (c = 1.0, $CHCl_3$). HRMS-ESI-ORBITRAP (+): calculated for $C_{23}H_{29}NO_3$ 390.2045, found 390.2049 $[M+H]^+$. ^1H NMR (600 MHz, $CDCl_3$): δ 1.24 (s, 3H), 1.30 (s, 9H), 1.43 (s, 3H), 1.97 (m, 1H), 2.07 (m, 1H), 2.31 (m, 1H), 2.41 (m, 2H), 2.48 (m, 1H), 2.53 (dd, 1H, J_1 = 18.78 Hz, J_2 = 4.26 Hz), 2.86 (dd, 1H, J_1 = 18.76 Hz, J_2 = 9.81 Hz), 3.20 (dd, 1H, J_1 = 9.76 Hz, J_2 = 4.19 Hz), 5.98 (s, 1H), 6.76 (dd, 1H, J_1 = 7.82 Hz, J_2 = 1.30 Hz), 7.28 (m, 1H), 7.39 (m, 1H), 7.58 (dd, 1H, J_1 = 8.15 Hz, J_2 = 1.00 Hz). ^{13}CNMR (150 MHz, $CDCl_3$): δ 14.1, 21.7, 22.6, 23.1, 24.0, 26.0, 31.5, 31.6, 32.3, 35.7, 37.4, 42.9, 46.3, 125.5, 127.5, 129.0, 129.9, 130.0, 130.4, 147.9, 168.5, 175.9, 177.7, 199.6.

(P)-(R)-1-(2-(tert-butyl)-4-chlorophenyl)-3-(2-(3-oxocyclohex-1-en-1-yl)propan-2-yl)pyrrolidine-2,5-dione (product **163**, Table 3.3—entry 2)

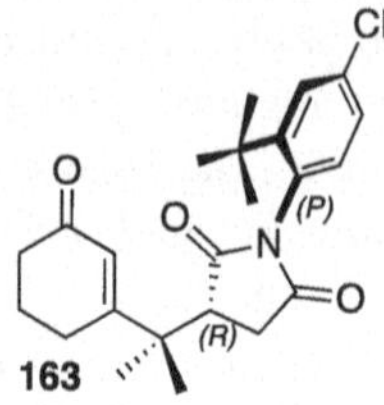

163

The reaction was carried at room temperature following the general procedure to furnish the crude product **163**. The crude mixture obtained has been purified by flash column chromatography (hexane/ethyl acetate = 3/2) in 75% yield and 99% ee. The ee was determined by HPLC analysis on a Daicel Chiralpak AD-H column: hexane/*i*-PrOH 80:20, flow rate 1.0 mL/min, $\lambda = 254$ nm: **163** $\tau_{major} = 8.9$ min, $\tau_{minor} = 12.7$ min. $[\alpha]_D^{20} = +55.6$ ($c = 1.0$, $CHCl_3$). HRMS-ESI-ORBITRAP(+): calculated for $C_{23}H_{28}ClNO_3$ 402.1830, found 402.1826 $[M+H]^+$. ^1H NMR (600 MHz, $CDCl_3$): δ 1.23 (s, 3H), 1.28 (s, 9H), 1.43 (s, 3H), 2.03 (m, 2H), 2.24–2.61 (m, 5H), 2.86 (dd, 1H, $J_1 = 18.6$ Hz, $J_2 = 9.5$ Hz), 3.20 (dd, 1H, $J_1 = 9.5$ Hz, $J_2 = 4.2$ Hz), 5.98 (s, 1H), 6.71 (d, 1H, $J = 8.1$ Hz), 7.26 (dd, 1H, $J_1 = 8.1$ Hz, $J_2 = 2.3$ Hz), 7.54 (d, 1H, $J = 2.3$ Hz). ^{13}C NMR (150 MHz, $CDCl_3$): δ 21.7, 23.1, 24.1, 26.0, 31.4, 32.3, 35.9, 37.4, 42.9, 46.4, 125.6, 127.7, 128.7, 129.4, 131.8, 135.8, 150.0, 168.2, 175.6, 177.5, 199.6.

(P)-(R)-1-(3-(tert-butyl)-[1,1′-biphenyl]-4-yl)-3-(2-(3-oxocyclohex-1-en-1-yl)propan-2-yl)pyrrolidine-2,5-dione (product **164**, Table 3.3—entry 3)

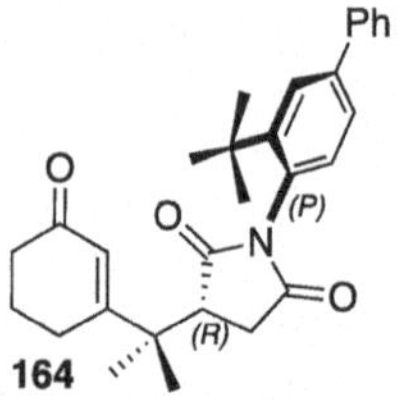

164

The reaction was carried at room temperature following the general procedure to furnish the crude product **164**. The crude mixture obtained has been purified by flash column chromatography (hexane/ethyl acetate = 3/2) in 68% yield and 97% ee. The ee was determined by HPLC analysis on a Daicel Chiralpak AD-H column: hexane/*i*-PrOH 80:20, flow rate 1.0 mL/min, $\lambda = 254$ nm: **164** $\tau_{major} = 11.1$ min, $\tau_{minor} = 22.2$ min. $[\alpha]_D^{20} = +62.0$ ($c = 1.0$, $CHCl_3$). HRMS-ESI-ORBITRAP(+): calculated for $C_{29}H_{33}NO_3$ 444.2533, found 444.2529 $[M+H]^+$. ^1H NMR (600 MHz, $CDCl_3$): δ 1.25 (s, 3H), 1.35 (s, 9H), 1.44 (s, 3H), 1.97 (m, 1H), 2.08 (m, 1H), 2.32 (m, 1H), 2.38–2.60 (m, 4H), 2.88 (dd, 1H, $J_1 = 18.9$ Hz, $J_2 = 9.81$ Hz), 3.22 (dd, 1H, $J_1 = 9.8$ Hz, $J_2 = 4.10$ Hz), 6.00 (s, 1H), 6.84 (d, 1H, $J_1 = 8.0$ Hz), 7.36 (m, 1H), 7.45 (m, 3H), 7.55 (m, 2H), 7.76 (d, 1H, $J_1 = 1.32$ Hz). ^{13}C NMR (150 MHz, $CDCl_3$): δ 21.7, 23.0, 24.0, 25.3, 25.9, 31.6, 32.3, 35.8, 37.4, 42.9, 46.3, 125.5, 126.3, 127.3, 127.6, 128.1, 128.7, 129.0, 130.7, 140.6, 142.8, 148.1, 168.4, 175.9, 177.8, 199.6.

(P)-(R)-1-(2-(tert-butyl)-4-methoxyphenyl)-3-(2-(3-oxocyclohex-1-en-1-yl)propan-2-yl)pyrrolidine-2,5-dione (product **165**, Table 3.3—entry 4)

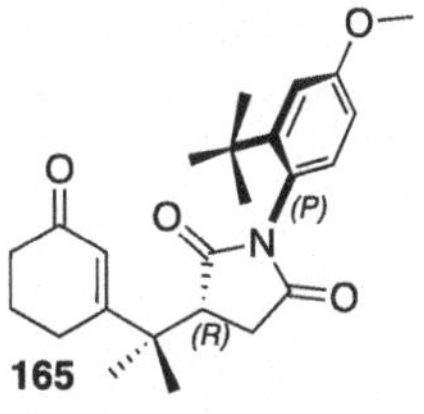

165

The reaction was carried at room temperature following the general procedure to furnish the crude product **165**. The crude mixture obtained has been purified by flash column chromatography (hexane/ethyl acetate = 3/2) in 74% yield and 98% ee. The ee was determined by HPLC analysis on a Daicel Chiralpak AD-H column: hexane/*i*-PrOH 80:20, flow rate 1.0 mL/min, λ = 254 nm: **165** τ_{major} = 12.7 min, τ_{minor} = 24.5 min. $[\alpha]_D^{20}$ = +55.2 (c = 1.0, $CHCl_3$). HRMS-ESI-ORBITRAP (+): calculated for $C_{24}H_{31}NO_4Na$ 420.2145, found 420.2136[M+Na]$^+$. ^{1}H NMR (600 MHz, $CDCl_3$): δ 1.23 (s, 3H), 1.27 (s, 9H), 1.42 (s, 3H), 1.97 (m, 1H), 2.08 (m, 1H), 2.31 (m, 1H), 2.41 (m, 2H), 2.51 (m, 2H), 2.85 (dd, 1H, J_1 = 18.6 Hz, J_2 = 9.7 Hz), 3.18 (dd, 1H, J_1 = 9.7 Hz, J_2 = 4.3 Hz), 3.80 (s, 3H), 5.98 (s, 1H), 6.70 (d, 1H, J = 8.6 Hz), 6.80 (dd, 1H, J_1 = 8.6 Hz, J_2 = 2.5 Hz), 7.09 (d, 1H, J_1 = 2.5 Hz). ^{13}C NMR (150 MHz, $CDCl_3$): δ 21.7, 23.1, 24.0, 26.0, 31.4, 32.2, 35.6, 37.4, 42.8, 46.1, 55.3, 111.6, 111.5, 122.6, 125.5, 131.4, 149.3, 160.1, 168.4, 176.1, 177.9, 199.6.

(P)-(R)-benzyl (4-(tert-butyl)-3-(2,5-dioxo-3-(2-(3-oxocyclohex-1-en-1-yl)propan-2-yl)pyrrolidin-1-yl)phenyl)carbamate (product **166**, Table 3.3—entry 5)

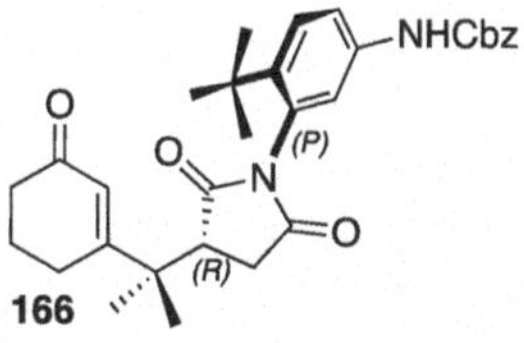

166

The reaction was carried at room temperature following the general procedure to furnish the crude product **166**. The crude mixture obtained has been purified by flash column chromatography (hexane/ethyl acetate = 3/2) in 81% yield and 97% ee. The ee was determined by HPLC analysis on a Daicel Chiralpak AD-H column: hexane/*i*-PrOH 70:30, flow rate 1.0 mL/min, λ = 254 nm: **166** τ_{major} = 24.1 min, τ_{minor} = 18.0 min. $[\alpha]_D^{20}$ = +43.0 (c = 1.0, $CHCl_3$). HRMS-ESI-ORBITRAP(+): calculated for $C_{31}H_{36}N_2O_5$ 517.2697, found 517.2682 [M+H]$^+$. ^{1}H NMR (600 MHz, $CDCl_3$): δ 1.17 (s, 3H), 1.25 (s, 9H), 1.37 (s, 3H), 1.92 (m, 1H), 2.03 (m, 1H), 2.26 (m, 1H), 2.34–2.51 (m, 4H), 2.81 (dd, 1H, J_1 = 18.5 Hz, J_2 = 9.5 Hz), 3.14 (dd, 1H, J_1 = 9.5 Hz, J_2 = 3.6 Hz), 5.13 (d, 1H, J = 12.2 Hz), 5.17 (d, 1H, J = 12.2 Hz),

5.93 (s, 1H), 7.05 (s, 1H), 7.2–7.44 (m, 8H). ^{13}C NMR (150 MHz, $CDCl_3$): δ 20.7, 22.1, 23.1, 24.3, 25.0, 30.6, 31.3, 34.2, 36.4, 42.0, 45.4, 65.9, 118.6, 124.5, 127.2, 127.3, 127.6, 128.4, 129.2, 135.0, 136.3, 141.2, 152.1, 167.8, 174.8, 176.8, 198.9.

(P)-(R)-1-(2-(tert-butyl)phenyl)-3-(1-(3-oxocyclohex-1-en-1-yl)cyclopentyl)pyrrolidine-2,5-dione (product **167**, Table 3.3—entry 6)

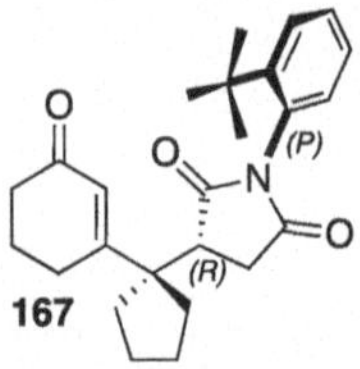

167

The reaction was carried at room temperature following the general procedure to furnish the crude product **167**. The crude mixture obtained has been purified by flash column chromatography (hexane/ethyl acetate = 3/2) in 80% yield and 97% ee. The ee was determined by HPLC analysis on a Daicel Chiralpak OD-H column: hexane/*i*-PrOH 80:20, flow rate 0.7 mL/min, λ = 254 nm: **167** τ_{major} = 18.7 min, τ_{minor} = 16.6 min. $[\alpha]_D^{20}$ = +45.5 (*c* = 1.0, $CHCl_3$). HRMS-ESI-ORBITRAP(+): calculated for $C_{25}H_{31}NO_3$ 394.2377, found 394.2369 $[M+H]^+$. ^{1}H NMR (600 MHz, $CDCl_3$): δ 1.29 (s, 9H), 1.6–2.1 (m, 10H), 2.32–2.50 (m, 4H), 2.60 (dd, 1H, J_1 = 18.5 Hz, J_2 = 3.9 Hz), 2.90 (dd, 1H, J_1 = 18.5 Hz, J_2 = 9.9 Hz), 3.31 (dd, 1H, J_1 = 9.9 Hz, J_2 = 3.9 Hz), 5.98 (s, 1H), 6.74 (d, 1H, J = 7.2 Hz), 7.27 (m, 1H), 7.39 (m, 1H), 7.58 (d, 1H, J = 8.3 Hz). ^{13}C NMR (150 MHz, $CDCl_3$): δ 23.3, 23.6, 23.8, 27.3, 31.5, 32.7, 33.1, 33.5, 35.6, 37.4, 45.0, 55.1, 126.3, 127.4, 129.0, 129.7, 129.9, 130.2, 147.8, 166.7, 175.8, 177.8, 199.4.

(P)-(R)-1-(2-(tert-butyl)phenyl)-3-((S)-2-(3-oxocyclohex-1-en-1-yl)butan-2-yl)pyrrolidine-2,5-dione and **(P)-(R)-1-(2-(tert-butyl)phenyl)-3-((R)-2-(3-oxocyclohex-1-en-1-yl)butan-2-yl)pyrrolidine-2,5-dione** (product **168a** and **168b**, Table 3.3—entry 7)

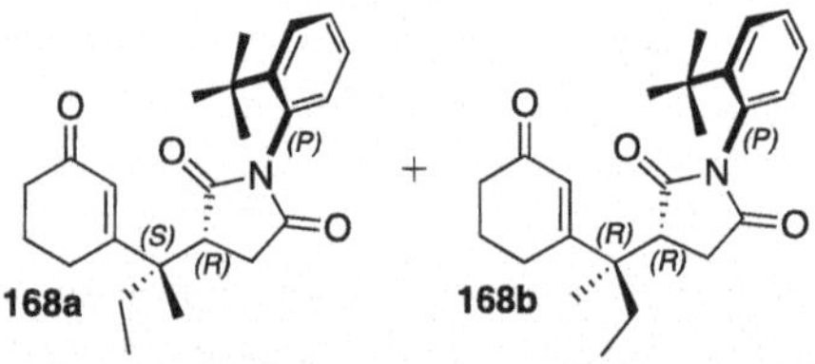

The reaction was carried out at room temperature following the general procedure to furnish the crude product as a 75:25 (^{1}H-NMR signal: δ_{major} 5.96 ppm. bs, δ_{minor} *5.97* ppm. bs) mixture of two diastereoisomers **168a** major and **168b** minor. The crude mixture obtained has been purified by flash column chromatography (hexane/ethyl acetate = 70/30) as pure **168a** and mixture of **168a** and **168a** isomers in 30% yield and 97% ee on **168a** and 95% ee on **168b**. The ee was determined by HPLC analysis on a Daicel Chiralpak OD-H column: hexane/*i*-PrOH 85:15, flow rate

0.5 mL/min, λ = 214 nm: **168a** τ_{major} = 52.2 min, τ_{minor} = 28.0 min; **168b** τ_{major} = 34.7 min, τ_{minor} = 30.9 min. $[\alpha]_D^{rt}$ = +69.1 (*c* = 1.0, $CHCl_3$, on pure **168b**). HRMS-ESIORBITRAP(+): calculated for $C_{24}H_{31}NO_3$ 382.2377, found 382.2369 $[M+H]^+$. ^{1}H-NMR (600 MHz, $CDCl_3$): (pure product **168a**) δ 0.76 (t, 1H, *J* = 7.5 Hz), 1.17 (s, 3H), 1.29 (s, 9H), 1.94–2.09 (m, 3H), 2.23 (m, 1H), 2.35 (m, 1H), 2.44 (m, 3H), 2.82 (dd, 1H, J_1 = 18.5 Hz, J_2 = 9.8 Hz), 3.14 (dd, 1H, J_1 = 9.8 Hz, J_2 = 4.2 Hz), 5.96 (s, 1H), 6.77 (dd, 1H, J_1 = 7.8 Hz, J_2 = 1.5 Hz), 7.28 (m, 1H), 7.40 (m, 1H), 7.59 (dd, 1H, J_1 = 8.2 Hz, J_2 = 1.5 Hz). ^{13}C NMR (150 MHz, $CDCl_3$): δ 8.7, 16.3, 22.9, 25.6, 29.5, 31.6, 32.5, 35.6, 37.4, 46.3, 47.3, 127.4, 128.0, 129.0, 129.8, 130.0, 130.3, 147.9, 165.9, 175.7, 177.9, 199.1. ^{1}H NMR (600 MHz, $CDCl_3$): (mixture of **168a**:**168b** = 2:1) δ 0.78 (m, 3.13H), 1.17 (m, 3.12H), 1.30 (m, 9.6H), 1.68–1.91 (m, 1.32H), 1.91–2.27 (m, 3.26), 2.27–2.59 (m, 5.10H), 2.80 (m, 1H), 2.94 (m, 0.32H), 3.13 (m, 0.64H), 3.28 (m, 0.32H), 5.97 (m, 1H), 6.75 (m, 1H), 7.28 (m, 1H), 7.39 (m, 1H), 7.57 (m, 1H). ^{13}C NMR (150 MHz, $CDCl_3$): δ 8.3, 8.8, 16.4, 17.1, 23.0, 25.7, 26.5, 29.6, 29.7, 30.2, 31.7, 32.6, 35.7, 35.8, 37.4, 37.5, 46.1, 46.4, 47.4, 47.6, 127.5, 127.6, 128.0, 129.0, 129.1, 129.8, 129.9, 130.1, 130.4, 130.5, 147.9, 148.0, 166.0, 166.7, 175.9, 176.1, 177.1, 178.0, 199.2, 199.7.

(P)-(R)-1-(4-bromo-2-(tert-butyl)phenyl)-3-((S)-2-(3-oxocyclohex-1-en-1-yl)butan-2-yl)pyrrolidine-2,5-dione and **(P)-(R)-1-(4-bromo-2-(tert-butyl)phenyl)-3-((R)-2-(3-oxocyclohex-1-en-1-yl)butan-2-yl)pyrrolidine-2,5-dione** (product **169a** and **169b**, Table 3.3—entry 8)

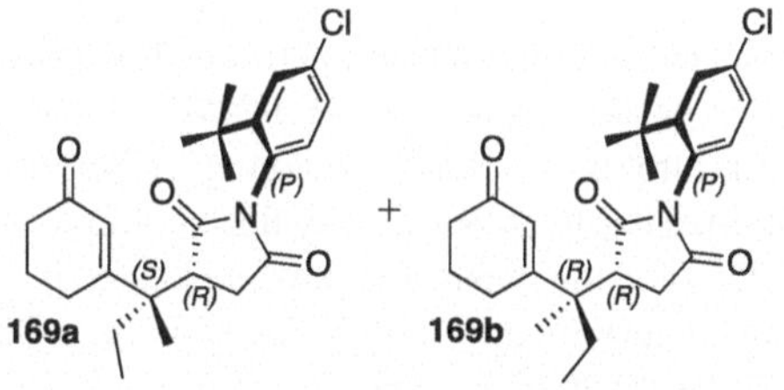

The reaction was carried out at room temperature following the general procedure to furnish the crude product as a 73:27 (1H-NMR signal: δ_{major} 5.95 ppm. bs, δ_{minor} 5.96 ppm. bs) mixture of two diastereoisomers **169a** major and **169b** minor. The crude mixture obtained has been purified by flash column chromatography (hexane/ethyl acetate = 7/3) as pure **169a** and mixture of **169a** and **169b** isomers in 35% yield and 98% ee on **169a** and 96% ee on **169b**. The ee was determined by HPLC analysis on a Daicel Chiralpak OD-H column: hexane/*i*-PrOH 80:20, flow rate 1.0 mL/min, λ = 214 nm: **169a** τ_{major} = 26.0 min, τ_{minor} = 11.3 min; **169b** τ_{major} = 15.7 min, τ_{minor} = 9.1 min. $[\alpha]_{rt}^{D}$ = +79.8 (*c* = 0.5, $CHCl_3$, on pure **169a**). HRMS-ESIORBITRAP(+): calculated for $C_{24}H_{30}BrNO_3$ 460.1482/462.1465, found 460.1472/462.1452 $[M+H]^+$. ^{1}H-NMR (600 MHz, $CDCl_3$): (pure product **169a**) δ 0.76 (t, 1H, *J* = 7.5 Hz), 1.15 (s, 3H), 1.28 (s, 9H), 1.84 (m, 1H), 2.02 (m, 2H), 2.21 (m, 1H), 2.34 (m, 2H), 2.44 (m, 3H), 2.81 (dd, 1H, J_1 – 18.7 Hz, J_2 = 9.6 Hz), 3.13 (dd, 1H, J_1 = 9.6 Hz, J_2 = 4.2 Hz), 5.95 (s, 1H), 6.64 (d, 1H, *J* = 8.2 Hz), 7.41

(dd, 1H, J_1 = 8.2 Hz, J_2 = 2.2 Hz), 7.70 (d, 1H, J = 2.2 Hz). ^{13}C NMR (150 MHz, $CDCl_3$): δ 8.7, 16.4, 23.0, 25.4, 29.6, 31.4, 32.5, 35.9, 37.5, 46.4, 47.3, 124.2, 128.1, 129.3, 130.7, 132.0, 132.4, 150.3, 165.7, 175.6, 177.7, 199.1. ^{1}H NMR (600 MHz, $CDCl_3$): (mixture of **169a**:**69b** = 70:30) δ 0.79 (m, 3.25H), 1.14 (m, 3.17), 1.28 (m, 9.57H), 1.60–2.14 (m, 3.80H), 2.14–2.53 (m, 6.10H), 2.84 (m, 1.30H), 3.14 (m, 0.75H), 3.28 (m, 0.25H), 5.95 (m, 1H), 6.63 (m, 1H), 7.40 (m, 1H), 7.69 (m, 1H). ^{13}C NMR (150 MHz, $CDCl_3$): δ 8.3, 8.7, 16.4, 17.1, 23.0, 25.6, 26.5, 29.6, 29.7, 30.1, 31.4, 31.7, 32.5, 35.9, 37.4, 37.5, 46.1, 46.4, 47.3, 47.6, 53.4, 124.2, 127.6, 128.1, 129.1, 129.3, 130.6, 132.0, 132.3, 132.4, 150.2, 150.3, 165.7, 166.5, 175.6, 175.7, 176.8, 177.7, 199.1, 199.6.

(P)-(R)-1-(2-(tert-butyl)phenyl)-3-((S)-2-(3-oxocyclohex-1-en-1-yl)-1-phenylpropan-2-yl)pyrrolidine-2,5-dione and **(P)-(R)-1-(2-(tert-butyl)phenyl)-3-((R)-2-(3-oxocyclohex-1-en-1-yl)-1-phenylpropan-2-yl)pyrrolidine-2,5-dione** (product **170a** and **170b**, Table 3.3—entry 9)

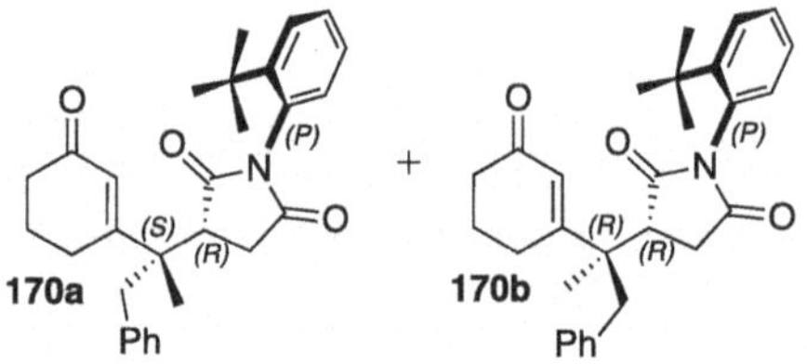

The reaction was carried out at room temperature following the general procedure to furnish the crude product as a 83:17 (1H-NMR signal: δ_{major} 5.66 ppm. bs, δ_{minor} 5.82 ppm. bs) mixture of two diastereoisomers **170a** major and **170b** minor. The crude mixture obtained has been purified by flash column chromatography (hexane/ethyl acetate = 75/25) and the two isomers were isolated in 51% yield and > 99% ee on **170a**. The ee was determined by HPLC analysis on a Daicel Chiralpak AD-H column: hexane/*i*-PrOH 80:20, flow rate 1.0 mL/min, λ = 254 nm: **170a** τ_{major} = 12.4 min, τ_{minor} = 18.4 min. It was not possible to obtain a pure enough representative amount of the minor diastereoisomer when performing the reaction with the pseudoenantiomer of catalyst **140** so the ee of **170b** could not be determined. $[\alpha]_D^{rt}$ = +71.3 (c = 1.0, $CHCl_3$, on pure **170a**). HRMS-ESIORBITRAP(+): calculated for $C_{29}H_{33}NO_3$ 444.2533, found 444.2529 $[M+H]^+$. ^{1}H NMR (600 MHz, $CDCl_3$): (pure product **170a**) δ 1.18 (s, 3H), 1.33 (s, 9H), 2.03 (m, 2H), 2.15 (m, 1H), 2.43 (m, 4H), 2.90 (dd, 1H, J_1 = 18.8 Hz, J_2 = 9.7 Hz), 3.25 (d, 1H, J = 13.8 Hz), 3.31 (dd, 1H, J_1 = 9.7 Hz, J_2 = 4.0 Hz), 3.54 (d, 1H, J = 13.8 Hz), 5.66 (s, 1H), 6.82 (m, 1H), 7.07 (m, 2H), 7.22 (m, 3H), 7.31 (m, 1H), 7.42 (m, 1H), 7.61 (m, 1H). ^{13}C NMR (150 MHz, $CDCl_3$): δ 17.5, 22.6, 26.7, 31.7, 33.0, 35.8, 37.3, 43.3, 45.5, 48.1, 126.9, 127.5, 128.1, 128.5, 129.1, 129.9, 130.0, 130.3, 130.4, 136.5, 148.0, 165.2, 175.7, 178.3, 198.9.

3.4.4 General Procedure for the Desymmetrization of Maleimides with Different Nucleophiles

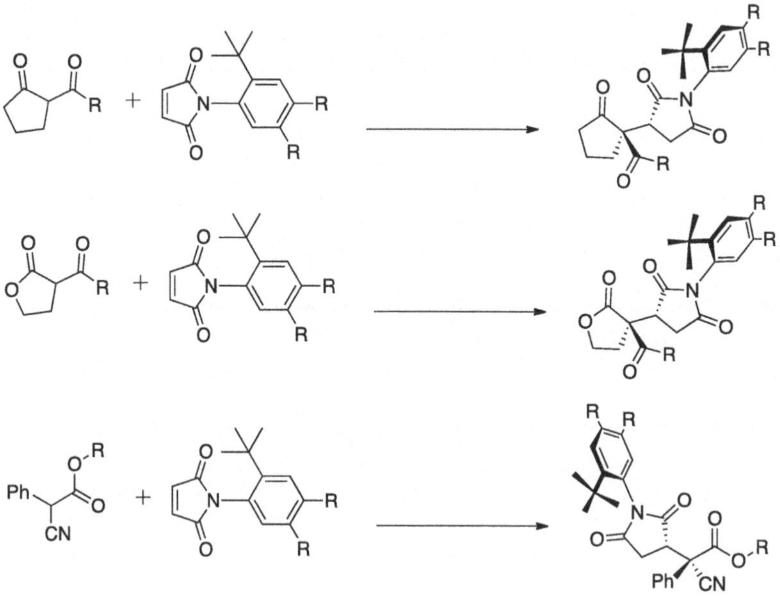

All the reactions were carried out in undistilled solvents and stirring was provided by magnetic Teflon-coated stir bars. In an ordinary vial containing the Michael donor (0.2 mmol, 1.0 equiv.), dichloromethane or acetone (0.8 mL; 0.25 M) and maleimide (0.2 mmol, 1.0 equiv.) were added and the vial was placed in a cold bath (previously set to −78 °C) for 10 min. At this point the vial was removed from the bath for a moment and catalyst **171**(10 mol%, 0.02 mmol) was quickly added before putting the vial back at −78 °C under magnetic stirring for 72 h. The crude mixture was then flushed through a short plug of silica, using dichloromethane/ethyl acetate 1:1 as the eluent (50 ml). Then solvent was removed in *vacuo* and the diastereomeric ratio (dr) was determined by ^{1}H NMR analysis of the crude mixture. Finally the desired compound was isolated by flash column chromatography and the enantiomeric excess determined by means of chiral HPLC analysis.

(*P*)-(*R*)-3-((*S*)-1-acetyl-2-oxocyclopentyl)-1-(2-(*tert*-butyl)phenyl)pyrrolidine-2,5-dione (product **173**, Table 3.4—entry 1)

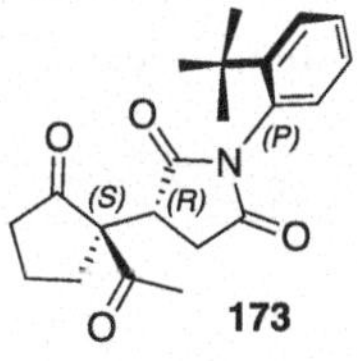

173

The title compound was obtained following the general procedure to furnish the crude product as a single diastereoisomer. The crude mixture was purified by flash column chromatography (hexane:ethyl acetate = 6:4) to give **173** in 85% yield and 93% ee. The ee was determined by HPLC analysis on a Daicel Chiralpak AD-H column: hexane/*i*-PrOH 80/20, flow rate 0.75 mL/min, λ = 214 nm: **173** τ_{major} = 13.1 min; τ_{minor} = 11.9 min. HRMS-ESI (+): calculated for $C_{21}H_{26}NO_4$ 356.1856, found 356.1854 $[M+H]^+$. ^1H NMR (400 MHz, $CDCl_3$) δ (ppm) 7.55 (1 H, dd, J_1 = 2.0 Hz, J_2 = 7.9 Hz), 7.38 (1 H, td, J_1 = 1.7 Hz, J_2 = 7.3 Hz, J_3 = 7.3 Hz), 7.30 (1 H, td, J_1 = 1.7 Hz, J_2 = 7.3 Hz, J_3 = 7.3 Hz), 7.09 (1 H, dd, J_1 = 1.7 Hz, J_2 = 7.7 Hz), 3.45 (1 H, q, J_1 = 5.8 Hz, J_2 = 9.7 Hz), 2.88 (1 H, q, J_1 = 9.7 Hz, J_2 = 18.6 Hz), 2.69–2.54 (3 H, m), 2.48 (2 H, t, J = 7.9 Hz), 2.37 (2 H, q), 2.20 (3 H, s), 1.28 (9 H, s). ^{13}C NMR: (100 MHz, $CDCl_3$) δ (ppm) 229.9, 216.4, 185.8, 183.7, 150.9, 130.6, 129.8, 128.5, 127.6, 58.14, 38.3, 35.2, 32.2, 31.8, 31.6, 30.5, 26.8, 19.6.

(*P*)-(*R*)-3-((*S*)-1-acetyl-2-oxocyclopentyl)-1-(4-bromo-2-(*tert*-butyl)phenyl)pyrrolidine-2,5-dione (product **174**, Table 3.4—entry 2)

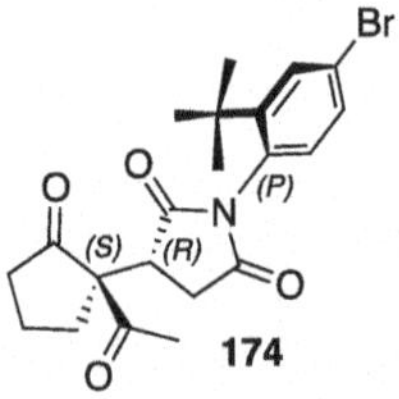

174

The title compound was obtained following the general procedure to furnish the crude product with a d.r. of 17:1. The crude mixture was purified by flash column chromatography (hexane:ethyl acetate = 6:4) to give **174** in 86% yield and 94% ee. The ee was determined by HPLC analysis on a Daicel Chiralpak AD-H column: hexane/*i*-PrOH 90/10, flow rate 0.75 mL/min, λ = 214 nm: **174** τ_{major} = 41.19 min; τ_{minor} = 24.90 min. HRMS-ESI (+): calculated for $C_{21}H_{25}BrNO_4$ 434.0961, found 434.0961 $[M+H]^+$. ^1H NMR (300 MHz, $CDCl_3$): δ (ppm) 7.66 (d, J_1 = 2.2 Hz, 1H), 7.44 (dd, J_1 = 8.3 Hz, J_2 = 2.2 Hz, 1H), 7.03 (d, J_1 = 8.3 Hz, 1H), 3.35 (dd, J_1 = 9.6 Hz, J_2 = 5.8 Hz, 1H), 2.86 (dd, J_1 = 18.7 Hz, J_2 = 9.6 Hz, 1H), 2.71 (dd, J_1 = 18.7 Hz, J_2 = 5.8 Hz, 1H), 2.59–2.42 (m, 4H), 2.20 (s, 3H), 2.10 (m, 2H), 1.27 (s, 9H). ^{13}C NMR (75 MHz, $CDCl_3$): δ (ppm) 214.4 (C), 203.6 (C), 177.4 (C), 175.6 (C), 150.2 (C), 132.2 (CH), 131.8 (CH), 130.8 (CH), 129.9 (C), 124.1 (C), 69.0 (C), 42.1 (CH), 38.3 (CH_2), 35.7 (C), 32.2 (CH_2), 31.4 $(CH_3)_3$, 31.1 (CH_2), 26.9 (CH_3), 19.6 (CH_2).

(*P*)-(*R*)-3-((*S*)-1-acetyl-2-oxocyclopentyl)-1-(2-(*tert*-butyl)-4-chlorophenyl)pyrrolidine-2,5-dione (product **175**, Table 3.4—entry 3)

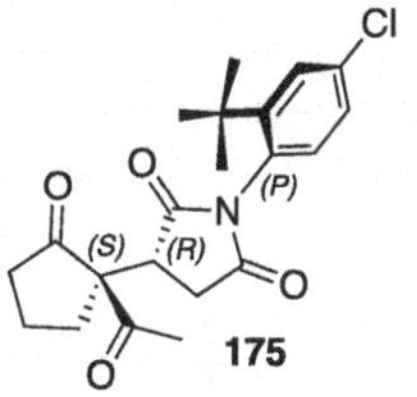

175

The title compound was obtained following the general procedure to furnish the crude product as a single diastereoisomer. The crude mixture was purified by flash column chromatography (hexane:ethyl acetate = 6:4) to give **175** in 81% yield and 93% ee. The ee was determined by HPLC analysis on a Daicel Chiralpak AD-H column: hexane/*i*-PrOH 80/20, flow rate 1.0 mL/min, λ = 214 nm: **175** τ_{major} = 11.90 min; τ_{minor} = 8.28 min. $[\alpha]_D^{20}$ +2.5 (*c* 1.00, $CHCl_3$). HRMS-ESI (+): calculated for $C_{21}H_{25}ClNO_4$ 390.1467, found 390.1478 $[M+H]^+$. ^1H NMR (400 MHz, $CDCl_3$): δ (ppm): 7.51 (*d*, 1H, *J* = 2.3 Hz); 7.29 (*dd*, 1H, J_1 = 8.5 Hz, J_2 = 2.3 Hz); 7.10 (*d*, 1H, *J* = 8.5 Hz); 3.36 (*dd*, 1H, J_1 = 9.7 Hz, J_2 = 5.7 Hz); 2.86 (*dd*, 1H, J_1 = 18.8 Hz, J_2 = 9.7 Hz); 2.71 (*dd*, 1H, J_1 = 18.7 Hz, J_2 = 5.7 Hz); 2.60–2.45 (*m*, 4H); 2.20 (*s*, 3H); 2.10 (*m*, 2H); 1.27 (*s*, 9H). ^{13}C NMR (100 MHz, $CDCl_3$): δ (ppm): 214.4, 203.6, 177.5, 175.6, 150.0, 135.6, 132.0, 129.3, 128.8, 127.7, 69.0, 42.1, 38.3, 35.7, 32.2, 31.3, 31.1, 26.8, 19.6.

***(P)*-*(R)*-3-(*(S)*-1-acetyl-2-oxocyclopentyl)-1-(2-(*tert*-butyl)-5-nitrophenyl)pyrrolidine-2,5-dione** (product **176**, Table 3.4—entry 4)

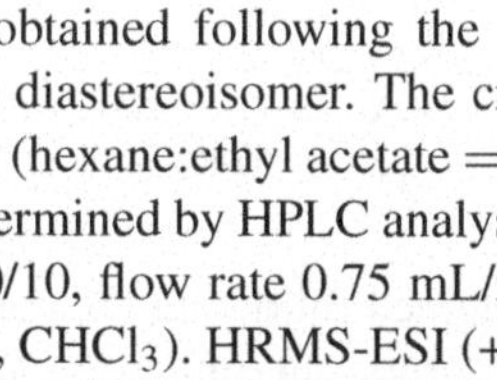

176

The title compound was obtained following the general procedure to furnish the crude product as a single diastereoisomer. The crude mixture was purified by flash column chromatography (hexane:ethyl acetate = 1:1) to give **176** in 82% yield and > 99% ee. The ee was determined by HPLC analysis on a Daicel Chiralpak AD-H column: hexane/*i*-PrOH 90/10, flow rate 0.75 mL/min, λ = 214 nm: **176** τ_{major} = 51.98 min. $[\alpha]_D^{20}$5.0 (*c* 1.00, $CHCl_3$). HRMS-ESI (+): calculated for $C_{21}H_{25}N_2O_6$ 401.1707, found 401.1714 $[M+H]^+$. ^1H NMR (300 MHz, $CDCl_3$): δ (ppm) 8.22 (dd, J_1 = 2.6 Hz, J_2 = 9.0 Hz, 1H); 8.10 (d, *J* = 2.6 Hz, 1H); 7.75 (d, *J* = 9.0 Hz, 1H); 3.37 (dd, J_1 = 6.0 Hz, J_2 = 9.3 Hz, 1H); 2.91 (dd, J_1 = 9.3 Hz, J_2 = 18.7 Hz, 1H); 2.79 (dd, J_1 = 6.0 Hz, J_2 = 18.7 Hz, 1H); 2.50 (m, 4H); 2.23 (s, 3H); 2.16 (m, 2H); 1.33 (s, 9H). ^{13}C NMR (75 MHz, $CDCl_3$): δ (ppm) 214.5 (C), 203.7 (C), 177.3 (C), 175.4 (C), 156.1 (C), 146,7 (C), 132.0 (C) 129.7 (CH), 126.3 (CH), 124.2 (CH), 69.1 (C), 42.1 (CH) 38.3 (CH_2), 36.4 (C), 32.2 (CH_2), 31.4 $(CH_3)_3$, 26.9 (CH_3) 19.7 (CH_2).

(*P*)-(*R*)-3-((*S*)-1-acetyl-2-oxocyclopentyl)-1-(2-(*tert*-butyl)-4-methoxyphenyl)pyrrolidine-2,5-dione (product **177**, Table 3.4—entry 5)

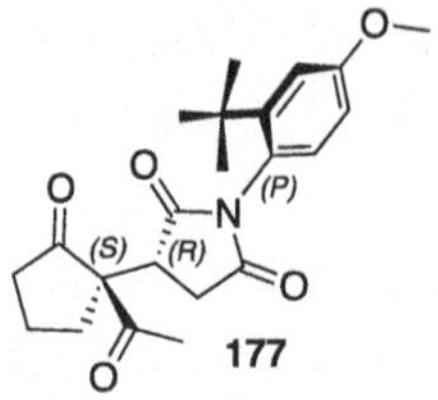

177

The title compound was obtained following the general procedure to furnish the crude with a d.r. of 19:1. The crude mixture was purified by flash column chromatography (hexane:ethyl acetate = 1:1) to give **177** in 90% yield and 94% ee. The ee was determined by HPLC analysis on a Daicel Chiralpak AD-H column: hexane/*i*-PrOH 80/20, flow rate 0.75 mL/min, $\lambda = 214$ nm: **177** $\tau_{major} = 27.7$ min; $\tau_{minor} = 17.4$ min. $[\alpha]_D^{20}$ +8.0 (*c* 1.00, $CHCl_3$). HRMS-ESI (+): calculated for $C_{22}H_{28}NO_5$ 386.1962, found 386.1963 $[M+H]^+$. ^1H NMR (300 MHz, $CDCl_3$) δ (ppm) 7.06 (d, $J = 2.8$ Hz, 1H), 7.01 (d, $J = 8.6$ Hz, 1H), 6.82 (dd, $J_1 = 8.6$ Hz, $J_2 = 2.8$ Hz, 1H), 3.80 (s, 3H), 3.44 (dd, $J_1 = 9.7$ Hz, $J_2 = 5.8$ Hz, 1H), 2.86 (dd, $J_1 = 18.6$ Hz, $J_2 = 9.7$ Hz, 1H), 2.63 (dd, $J_1 = 18.6$ Hz, $J_2 = 5.8$ Hz, 1H), 2.69–2.43 (m, 3H), 2.42–2.28 (m, 1H), 2.20 (s, 3H), 2.16–1.96 (m, 2H), 1.25 (s, 9H). ^{13}C NMR (100 MHz, $CDCl_3$): δ (ppm) 214.2 (C), 203.2 (C), 177.7 (C), 176.0 (C), 160.1 (C), 149.3 (C), 131.6 (CH), 123.1 (C), 115.1 (CH), 111.7 (CH), 69.0 (C), 55.3 (CH_3), 42.6 (CH), 38.32 (CH_2), 35.5 (C), 32.1 (CH_2), 31.4 $(CH_3)_3$, 30.5 (CH_2), 26.8 (CH_3), 19.6 (CH_2).

(*P*)-(*R*)-3-((*S*)-1-acetyl-2-oxocyclopentyl)-1-(2,5-di-*tert*-butylphenyl)pyrrolidine-2,5-dione (product **178**, Table 3.4—entry 6)

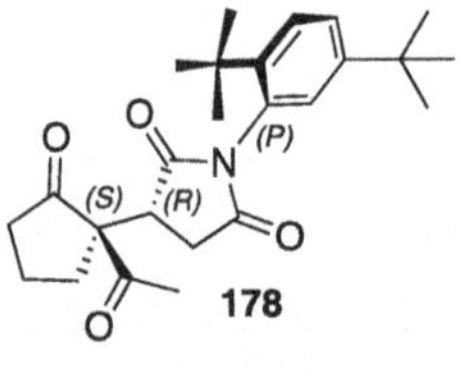

178

The title compound was obtained following the general procedure to furnish the crude with a d.r. of 18:1. The crude mixture was purified by flash column chromatography (hexane:ethyl acetate = 6:4) to give **178** in 85% yield and 87% ee. The ee was determined by HPLC analysis on a Daicel Chiralpak AS-H column: hexane/*i*-PrOH 95/5, flow rate 0.75 mL/min, $\lambda = 214$ nm: **178** $\tau_{major} = 30.37$ min; $\tau_{minor} = 37.19$ min. $[\alpha]_D^{20}$ +14.6 (*c* 1.00, $CHCl_3$). HRMS-ESI (+): calculated for $C_{25}H_{34}NO_4$ 412.2482, found 412.2479 $[M+H]^+$. ^1H NMR (400 MHz, $CDCl_3$) δ (ppm) 7.46 (1H, d, $J = 8.6$ Hz), 7.38 (1H, dd, $J_1 = 2.2$ Hz, $J_2 = 8.6$ Hz), 7.02 (1H, d, $J = 2.2$ Hz), 3.47 (1H, q, $J_1 = 5.8$ Hz, $J_2 = 9.7$ Hz), 2.89 (1H, q, $J_1 = 9.7$ Hz, $J_2 = 18.5$ Hz), 2.68–2.56 (3H, m), 2.49 (2H, t, $J = 7.7$ Hz), 2.39–2.29 (2H, m), 2.22

(3H, s), 1.30 (9H, s), 1.26 (9H, s). ^{13}C-NMR: (100 MHz, $CDCl_3$) δ (ppm) 214.1 (C), 177.5 (C), 175.9 (C), 150.4 (C), 144.5 (C), 128.2 (CH), 127.2 (CH), 126.9 (CH), 69.0 (C), 42.6 (CH) 38.3 (CH_2), 35.1 (C), 32.3 (CH_2), 31.6 (CH), 31.1 (CH), 30.5 (CH_2), 26.7 (CH), 19.6 (CH_2).

(*P*)-(*R*)-3-((*S*)-1-acetyl-2-oxocyclopentyl)-1-(4-bromo-2,5-di-*tert*-butylphenyl)pyrrolidine-2,5-dione (product **179**, Table 3.4—entry 7)

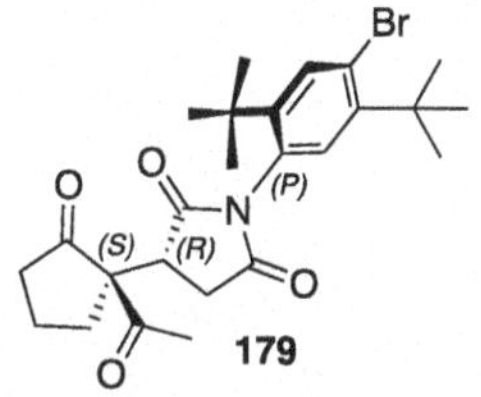

179

The title compound was obtained following the general procedure as single diastereoisomer. The crude mixture was purified by flash column chromatography (hexane:ethyl acetate = 6:4) to give **179** in 55% yield and 79% ee. The ee was determined by HPLC analysis on a Daicel Chiralpak AD-H column: hexane/*i*-PrOH 80/20, flow rate 0.75 mL/min, λ = 214 nm: **179** τ_{major} = 6.89 min; τ_{minor} = 7.74 min. HRMS-ESI (+): calculated for $C_{25}H_{33}BrNO_4$ 490.1587, found 490.1596 $[M+H]^+$. ^{1}H NMR (300 MHz, $CDCl_3$): δ (ppm) 7.71 (s, 1H); 7.16 (s, 1H); 3.36 (dd, J_1 = 5.7 Hz, J_2 = 10.0 Hz, 1H); 2.86 (dd, J_1 = 10.0 Hz, J_2 = 18.6 Hz, 1H); 2.70 (dd, J_1 = 5.7, J_2 = 18.6, 1H); 2.50 (m, 4H); 2.21 (s, 3H); 2.12 (m, 2H); 1.49 (s, 9H); 1.25 (s, 9H). ^{13}C NMR (75 MHz, $CDCl_3$): δ (ppm) 214.4 (C), 203.4 (C), 177.4 (C), 175.7 (C), 147.2 (C), 146.8 (C), 135.7 (CH), 130.2 (CH), 129.5 (C), 124.1 (C), 69.0 (C), 42.2 (CH), 38.3 (CH_2), 36.2 (C), 35.0 (C), 32.2 (CH_2), 31.3 $(CH_3)_3$, 31.0 (CH_2), 29.4 $(CH_3)_3$, 26.9 (CH_3), 19.6 (CH_2).

(*P*)-(*R*)-1-(2-(*tert*-butyl)phenyl)-3-((*S*)-2-oxo-1-propionylcyclopentyl)pyrrolidine-2,5-dione (product **180**, Table 3.4—entry 8)

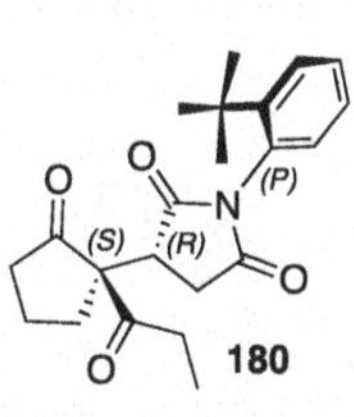

180

The title compound was obtained following the general procedure to furnish the crude with a d.r. of 9:1. The crude mixture was purified by flash column chromatography (hexane:ethyl acetate = 6:4) to give **180** in 65% yield and 97% ee. The ee was determined by HPLC analysis on a Daicel Chiralpak AD-H column: hexane/*i*-PrOH 80/20, flow rate 0.75 mL/min, λ = 214 nm: **180** τ_{major} = 10.43 min; τ_{minor} = 9.18 min. HRMS-ESI (+): calculated for $C_{22}H_{28}NO_4$ 370.2013, found 370.2013 $[M+H]^+$. ^{1}H

NMR (300 MHz, $CDCl_3$): δ (ppm) 7.55 (1H, dd, $J_1 = 1.6$ Hz, $J_2 = 8.0$ Hz), 7.43–7.29 (2H, m), 7.13 (1H, dd, $J_1 = 1.6$ Hz, $J_2 = 7.6$ Hz), 3.46 (1H, dd, $J_1 = 5.8$ Hz, $J_2 = 10.0$ Hz), 2.89 (1H, dd, $J_1 = 9.9$ Hz, $J_2 = 18.7$ Hz), 2.70–2.30 (8H, m), 2.16–1.98 (2H, m), 1.28 (9H, s), 1.08 (3H, t, $J = 9.9$ Hz). ^{13}C NMR (75 MHz, $CDCl_3$): δ (ppm) 214.7 206.1, 177.6, 175.8, 147.8, 130.6, 130.5, 129.7, 128.5, 127.6, 68.7, 42.9, 38.4, 35.5, 32.9, 32.4, 31.6, 31.1, 19.7.

(*P*)-(*R*)-1-(2-(*tert*-butyl)phenyl)-3-((*R*)-2-oxo-1-(2-phenylacetyl)cyclopentyl)pyrrolidine-2,5-dione (product **181**, Table 3.4—entry 9)

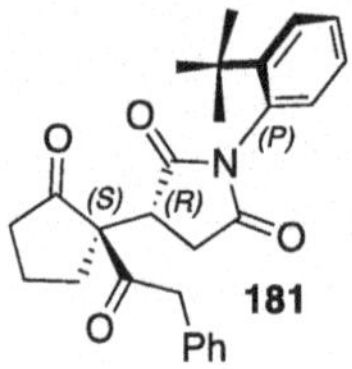

181

The title compound was obtained following the general procedure to furnish the crude with a d.r. of 4:1. The crude mixture was purified by flash column chromatography (hexane:ethyl acetate = 7:3) to give **181** in 36% yield and 37% ee. The ee was determined by HPLC analysis on a Daicel Chiralpak AD-H column: hexane/*i*-PrOH 80/20, flow rate 0.75 mL/min, $\lambda = 214$ nm: **181** $\tau_{major} = 13.54$ min; $\tau_{minor} = 10.96$ min. HRMS-ESI (+): calculated for $C_{27}H_{30}NO_4$ 432.2169, found 432.2172 $[M+H]^+$. ^1H NMR (300 MHz, $CDCl_3$): δ (ppm) 7.55 (1H, dd, $J_1 = 1.5$ Hz, $J_2 = 8.0$ Hz), 7.41–7.27 (5H, m), 7.18 (2H, dd, $J_1 = 1.3$ Hz, $J_2 = 8.4$ Hz), 7.06 (1H, dd, $J_1 = 1.6$ Hz, $J_2 = 7.7$ Hz), 3.86 (2H, s), 3.58 (1H, dd, $J_1 = 6.0$ Hz, $J_2 = 9.6$ Hz), 2.77 (1H, dd, $J_1 = 9.7$ Hz, $J_2 = 18.7$ Hz), 2.82–2.43 (5H, m), 2.33–2.23 (2 H, m), 1.28 (9H, s). ^{13}C NMR (150 MHz, $CDCl_3$): δ (ppm) 19.2, 29.1, 31.6, 32.8, 35.6, 38.3, 42.8, 45.2, 69.3, 127.3, 127.6, 128.6, 128.7, 129.8, 129.9, 130.0, 130.7, 132.9, 147.8, 175.3, 177.6, 201.3, 214.9.

(*P*)-Ethyl (*R*)-1-((*R*)-1-(2-(*tert*-butyl)phenyl)-2,5-dioxopyrrolidin-3-yl)-2-oxocyclopentane-1-carboxylate (product **182**, Table 3.4—entry 10)

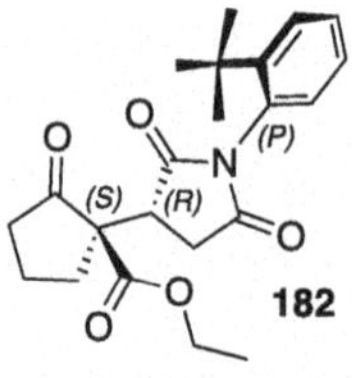

182

The title compound was obtained following the general procedure to furnish the crude with a d.r. of 8:1. The crude mixture was purified by flash column chromatography (hexane:ethyl acetate = 7:3) to give **182** in 50% yield and 50% ee. The ee was determined by HPLC analysis on a Daicel Chiralpak AD-H column: hexane/*i*-PrOH 80/20, flow rate 0.75 mL/min, $\lambda = 214$ nm: **182** $\tau_{major} = 12.7$ min; τ_{minor}

= 9.3 min. HRMS-ESI (+): calculated for $C_{22}H_{28}NO_5$ 386.1962, found 386.1960 $[M+H]^+$. ^{1}HNMR: (300 MHz, $CDCl_3$): δ (ppm) 7.55 (1H, dd, J_1 = 1.6 Hz, J_2 = 8.2 Hz), 7.38 (1H, td, J_1 = 1.6 Hz, J_2 = 7.5 Hz), 7.30 (1H, td, J_1 = 1.6 Hz, J_2 = 7.5 Hz), 7.04 (1H, dd, J_1 = 1.6 Hz, J_2 = 7.5 Hz), 3.41 (1H, q, J_1 = 5.9 Hz, J_2 = 9.6 Hz), 3.05 (1H, q, J_1 = 9.7 Hz, J_2 = 18.9 Hz), 2.87 (1H, dd, J_1 = 5.7 Hz, J_2 = 18.7 Hz), 2.68–2.51 (2H, m), 2.47–2.33 (2H, m), 2.19–2.04 (2H, m), 1.25 (9 H, s). ^{13}CNMR: (75 MHz, $CDCl_3$): δ (ppm) 214.7, 178.0, 175.7, 170.3, 147.8, 134.0, 131.4, 130.8, 130.4, 129.9, 129.8, 128.7, 128.4, 127.6, 127.3, 62.2, 61.3, 42.1, 38.0, 35.5, 33.8, 33.1, 31.6, 31.5, 19.2, 14.1.

(*P*)-(*R*)-3-((*S*)-3-acetyl-2-oxotetrahydrofuran-3-yl)-1-(2-(tert-butyl)phenyl)pyrrolidine-2,5-dione (product **183**, Table 3.5—entry 1)

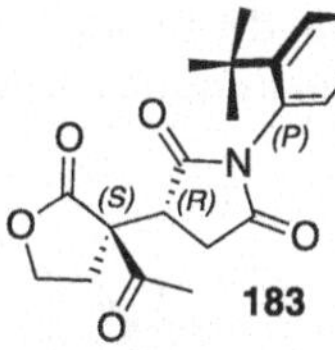

183

The title compound was obtained following the general procedure to furnish the crude product as a single diastereoisomer. The crude mixture was purified by flash column chromatography (hexane:ethyl acetate = 1:1) to give **183** in 85% yield and 99% ee. The ee was determined by HPLC analysis on a Daicel Chiralpak AD-H column: hexane/*i*-PrOH 80/20, flow rate 0.75 mL/min, λ = 214 nm: **183** τ_{major} = 14.04 min; τ_{minor} = 12.87 min. $[\alpha]_D^{20}$ −0.8 (*c* 1.00, $CHCl_3$). HRMS-ESI (+): calculated for $C_{20}H_{24}NO_5$ 358.1649, found 358.1646 $[M+H]^+$. ^{1}H NMR: (300 MHz, $CDCl_3$): δ (ppm) 7.56 (1H, dd, J_1 = 1.7 Hz, J_2 = 8.0 Hz), 7.39 (1H, td, J_1 = 1.7 Hz, J_2 = 7.2 Hz, J_3 = 7.2 Hz), 7.32 (1H, td, J_1 = 1.7 Hz, J_2 = 7.6 Hz, J_3 = 7.6 Hz), 7.19 (1H, dd, J_1 = 1.6 Hz, J_2 = 7.6 Hz), 4.73 (2H, t, J_1 = 7.3 Hz), 3.32 (1H, q, J_1 = 5.9 Hz, J_2 = 9.3 Hz), 3.04–2.88 (3H, m), 2.84–2.73 (1H, m), 2.33 (3H, s), 1.29 (9H, s). ^{13}C NMR: (75 MHz, $CDCl_3$): δ (ppm) 201.7, 177.1, 175.5, 174.0, 147.8, 130.6, 130.5, 129.8, 128.4, 127.6, 66.2, 61.9, 41.5, 35.5, 32.2, 31.6, 30.7, 26.4.

(*P*)-(*R*)-3-((*S*)-3-acetyl-2-oxotetrahydrofuran-3-yl)-1-(4-bromo-2-(tert-butyl)phenyl)pyrrolidine-2,5-dione (product **184**, Table 3.5—entry 2)

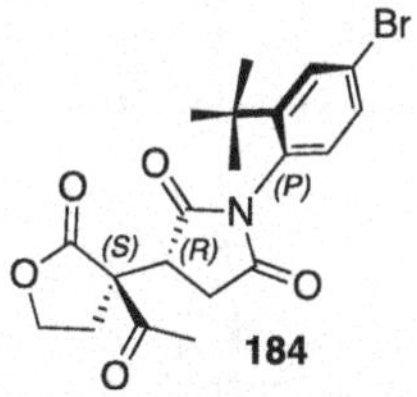

184

The title compound was obtained following the general procedure to furnish the crude product with a d.r. of 19:1. The crude mixture was purified by flash column chromatography (hexane:ethyl acetate = 1:1) to give **184** in 63% yield and 98% ee. The ee was determined by HPLC analysis on a Daicel Chiralpak AD-H column: hexane/*i*-PrOH 90/10, flow rate 0.75 mL/min, λ = 214 nm: **184** τ_{major} = 17.57 min; τ_{minor} = 13.54 min. $[\alpha]_D^{20}$ −5.7 (*c* 1.00, $CHCl_3$). $[\alpha]_D^{20}$ −2.1 (*c* 1.00, $CHCl_3$). HRMS-ESI (+): calculated for $C_{20}H_{23}BrNO_5$ 438.0736, found 436.0758 $[M+H]^+$. ^1H NMR (400 MHz, $CDCl_3$): δ (ppm): 7.68 (d, J_1 = 2.1 Hz, 1H); 7.46 (dd, J_1 = 2.1 Hz, J_2 = 8.4 Hz, 1H); 7.11 (d, J_1 = 8.4 Hz, 1H); 4.52 (m, 2H); 3.27(dd, J_1 = 6.7 Hz, J_2 = 8.3 Hz, 1H); 3.07(m, 1H); 2.94 (m, 2H); 2.80 (ddd, J_1 = 6.7 Hz, J_2 = 7.8 Hz, J_3 = 14.8 Hz, 1H); 2.33 (s, 3H); 1.28 (s, 9H). ^{13}C NMR (75 MHz, $CDCl_3$): δ (ppm): 201.8 (C), 176.9 (C), 175.3 (C), 174.0 (C), 150.2 (C), 132.2 (CH), 131.7 (CH), 130.8 (CH), 129.9 (C), 124.1 (C), 66.3 (CH_2), 61.9 (C), 41.4 (CH), 35.7 (C), 32.2 (CH_2), 31.4 $(CH_3)_3$, 30.9 (CH_2), 26.4 (CH_3).

(*P*)-(*R*)-3-((*S*)-3-acetyl-2-oxotetrahydrofuran-3-yl)-1-(2-(tert-butyl)-4-chlorophenyl)pyrrolidine-2,5-dione (product **185**, Table 3.5—entry 3)

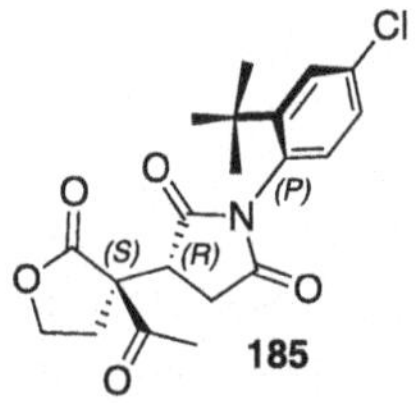

185

The title compound was obtained following the general procedure to furnish the crude product as a single diastereoisomer. The crude mixture was purified by flash column chromatography (hexane:ethyl acetate = 1:1) to give **185** in 93% yield and > 99% ee. The ee was determined by HPLC analysis on a Daicel Chiralpak AD-H column: hexane/*i*-PrOH 80/20, flow rate 1 mL/min, λ = 214 nm: **185** τ_{major} = 12.39 min; τ_{minor} = 9.76 min. $[\alpha]_D^{20}$ −5.7 (*c* 1.00, $CHCl_3$). HRMS-ESI (+): calculated for $C_{20}H_{23}ClNO_5$ 392.1259, found 392.1263 $[M+H]^+$. ^1H NMR (400 MHz, $CDCl_3$): δ (ppm): 7.52 (*d*, 1H, *J* = 2.3 Hz); 7.30 (*dd*, 1H, J_1 = 8.5 Hz, J_2 = 2.3 Hz); 7.17 (*d*, 1H, *J* = 8.5 Hz); 4.49 (*m*, 2H); 3.28 (*dd*, 1H, J_1 = 9.4 Hz, J_2 = 6.1 Hz); 3.10–2.85 (*m*, 3H); 2.77 (*m*, 1H); 2.32 (*s*, 3H); 1.28 (*s*, 9H). ^{13}C NMR (100 MHz, $CDCl_3$): δ (ppm): 202.0, 177.1, 175.5, 174.1, 150.1, 135.8, 132.0, 129.5, 128.9, 127.9, 66.4, 62.1, 41.5, 35.8, 32.3, 31.4, 31.0, 26.5.

(*P*)-(*R*)-3-((*S*)-3-acetyl-2-oxotetrahydrofuran-3-yl)-1-(2-(tert-butyl)-5-nitrophenyl)pyrrolidine-2,5-dione (product **186**, Table 3.5—entry 4)

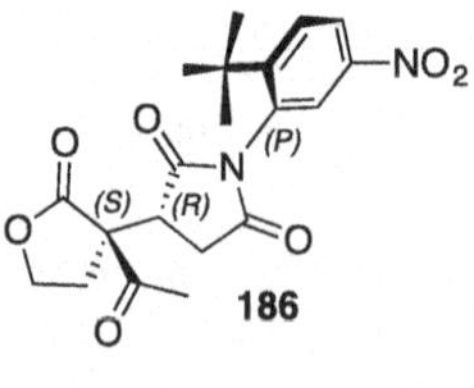

186

The title compound was obtained following the general procedure to furnish the crude with a d.r. of 7:1. The crude mixture was purified by flash column chromatography (hexane:ethyl acetate = 4:6) to give **186** in 69% yield and > 99% ee. The ee was determined by HPLC analysis on a Daicel Chiralpak AD-H column: hexane/*i*-PrOH 90/10, flow rate 0.75 mL/min, $\lambda = 214$ nm: **186** $\tau_{major} = 13.12$ min. $[\alpha]_D^{20} -19.2$ (*c* 1.00, $CHCl_3$). HRMS-ESI (+): calculated for $C_{20}H_{23}N_2O_7$ 403.150, found 403.1495 $[M+H]^+$. ^1H NMR (300 MHz, $CDCl_3$): δ (ppm): 8.19 (dd, $J_1 = 2.5$ Hz, $J_2 = 8.9$ Hz, 1H); 8.10 (d, $J_1 = 2.5$ Hz, 1H); 7.70 (d, $J_1 = 8.9$ Hz, 1H); 4.55 (m, 2H); 3.29 (m, 1H); 3.12 (ddd, $J_1 = 5.9$ Hz, $J_2 = 7.7$ Hz, $J_3 = 14.3$ Hz, 1H); 3.01 (m, 2H); 2.82 (ddd, $J_1 = 7.1$ Hz, $J_2 = 8.3$ Hz, $J_3 = 14.3$ Hz, 1H); 2.31 (s, 3H); 1.29 (s, 3H). ^{13}C NMR (75 MHz, $CDCl_3$): δ (ppm): 201.7 (C), 176.8 (C), 175.1 (C), 173.9 (C), 156.0 (C), 146.8 (C), 131.9 (C), 129.6 (CH), 126.2 (CH), 124.3 (CH), 66.3 (CH_2), 61.9 (C), 41.5 (CH), 36.4 (C), 32.3 (CH_2), 31.3 ($(CH_3)_3$), 30.9 (CH_2), 26.4 (CH_3).

(*P*)-(*R*)-3-((*S*)-3-acetyl-2-oxotetrahydrofuran-3-yl)-1-(2-(tert-butyl)-4-methoxyphenyl)pyrrolidine-2,5-dione (product **187**, Table 3.5—entry 5)

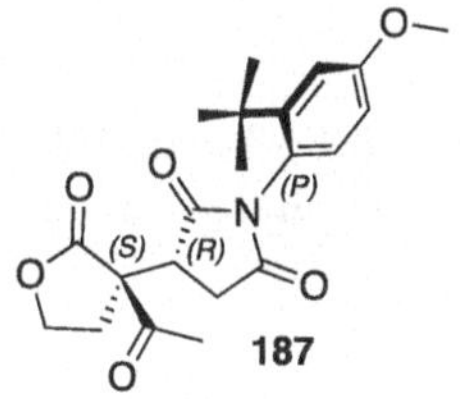

187

The title compound was obtained following the general procedure to furnish the crude with a d.r. of 19:1. The crude mixture was purified by flash column chromatography (hexane:ethyl acetate = 4:6) to give **187** in 67% yield and > 99% ee. The ee was determined by HPLC analysis on a Daicel Chiralpak AD-H column: hexane/*i*-PrOH 70/30, flow rate 0.75 mL/min, $\lambda = 214$ nm: **187** $\tau_{major} = 15.89$ min; $\tau_{minor} = 12.76$ min. $[\alpha]_D^{20} -0.5$ (*c* 1.00, $CHCl_3$). HRMS-ESI (+): calculated for $C_{21}H_{26}NO_6$ 388.1755, found 388.1756 $[M+H]^+$. ^1H NMR (300 MHz, $CDCl_3$): δ (ppm): 7.08 (d, $J_1 = 8.6$ Hz, 1H); 7.02 (d, $J_1 = 2.9$ Hz, 1H); 6.79 (dd, $J_1 = 2.9$ Hz, $J_2 = 8.6$ Hz, 1H); 4.44 (m, 2H); 3.77 (s, 3H); 3.26 (dd, $J_1 = 6.1$ Hz, $J_2 = 9.4$ Hz, 1H); 2.94 (dd, $J_1 = 7.1$ Hz, $J_2 = 13.9$ Hz, 1H); 2.86 (m, 2H); 2.75 (dd, $J_1 = 6.4$ Hz, $J_2 = 13.9$ Hz, 1H); 2.29 (s, 3H); 1.22 (s, 9H). ^{13}C NMR (75 MHz, $CDCl_3$): δ (ppm): 201.7 (C), 177.3 (C), 175.8 (C), 174.1 (C), 160.1 (C), 149.3 (C), 131.5 (CH), 123.2 (C), 115.1 (CH), 111.7 (CH), 66.2 (CH_2), 61.9 (C), 55.4 (CH_3), 41.4 (CH), 35.5 (C), 32.1 (CH_2), 31.4 ($(CH_3)_3$), 30.7 (CH_2), 26.5 (CH_3).

(*P*)-(*R*)-3-((*S*)-3-acetyl-2-oxotetrahydrofuran-3-yl)-1-(2,5-di-tert-butylphenyl)pyrrolidine-2,5-dione (product **188**, Table 3.5—entry 6)

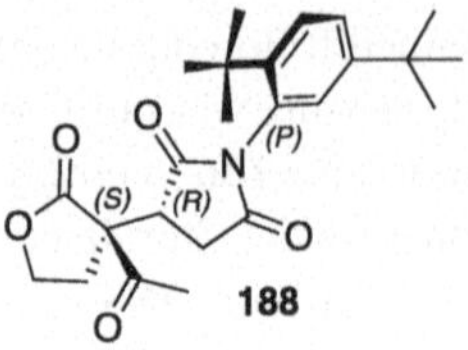

188

The title compound was obtained following the general procedure to furnish the crude with a d.r. of 6:1. The crude mixture was purified by flash column chromatography (hexane:ethyl acetate = 4:6) to give **188** in 71% yield and 90% ee. The ee was determined by HPLC analysis on a Daicel Chiralpak AD-H column: hexane/*i*-PrOH 95/5, flow rate 0.75 mL/min, λ = 214 nm: **188** τ_{major} = 25.03 min; τ_{minor} = 27.22 min. HRMS-ESI (+): calculated for $C_{24}H_{32}NO_5$ 414.2275, found 414.2283 $[M+H]^+$. ^{1}H NMR (400 MHz, $CDCl_3$): δ (ppm): 7.47 (1H, d, J = 8.5 Hz), 7.39 (1H, dd, J_1 = 2.3 Hz, J_2 = 8.6 Hz), 7.15 (1H, d, J = 2.3 Hz), 4.46 (2H, t, J = 7.2 Hz), 3.32 (1H, q, J_1 = 6.0 Hz, J_2 = 9.4 Hz), 3.02–2.87 (3H, m), 2.83–2.73 (1H, m), 2.33 (3H, s), 1.31 (9H, s), 1.27 (9H, s). ^{13}C NMR (100 MHz, $CDCl_3$): δ (ppm): 201.5, 177.1, 175.7, 174.1, 150.5, 144.5, 130.1, 128.0, 127.2, 126.8, 66.2, 62.0, 41.5, 35.0, 34.2, 32.2, 31.6, 31.1, 30.6, 26.5.

(*P*)-Benzyl (3-((*R*)-3-((*S*)-3-acetyl-2-oxotetrahydrofuran-3-yl)-2,5-dioxopyrrolidin-1-yl)-4-(tert-butyl)phenyl)carbamate (product **189**, Table 3.5—entry 7)

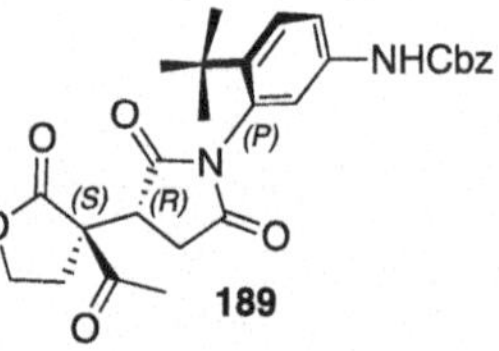

189

The title compound was obtained following the general procedure to furnish the crude product as a single diastereoisomer. The crude mixture was purified by flash column chromatography (hexane:ethyl acetate = 45:55) to give **189** in 90% yield and > 99% ee. The ee was determined by HPLC analysis on a Daicel Chiralpak AD-H column: hexane/*i*-PrOH 80/20, flow rate 1 mL/min, λ = 214 nm: **189** τ_{major} = 38.84 min. $[\alpha]_D^{20}$ −19.3 (*c* 2.00, $CHCl_3$). HRMS-ESI (+): calculated for $C_{28}H_{31}N_2O_7$ 507.2126, found 507.2123 $[M+H]^+$. ^{1}H NMR (400 MHz, $CDCl_3$): δ (ppm): 7.54 (*bs*, 1H); 7.42 (*d*, 1H, J = 8.7 Hz); 7.36–7.30 (*m*, 4H); 7.21 (*bs*, 1H); 7.06 (*d*, 1H, J = 2.5 Hz); 5.14 (*m*, 2H); 4.38 (*m*, 2H); 3.31 (*dd*, 1H, J_1 = 9.6 Hz, J_2 = 6.0 Hz); 2.98–2.77 (*m*, 3H); 2.67 (*m*, 1H); 2.26 (*s*, 3H); 1.23 (*s*, 9H). ^{13}C NMR(100 MHz, $CDCl_3$): δ (ppm): 201.8, 177.1, 175.7, 174.2, 153.3, 142.4, 137.5, 136.2, 130.7, 129.1, 128.6, 128.3, 128.2, 120.0, 119.9, 66.9, 66.4, 62.1, 41.7, 35.1, 32.3, 31.6, 30.4, 26.3.

(*P*)-(*R*)-3-((*S*)-3-(4-bromobenzoyl)-2-oxotetrahydrofuran-3-yl)-1-(2-(tert-butyl)phenyl)pyrrolidine-2,5-dione (product **190**, Table 3.5—entry 8)

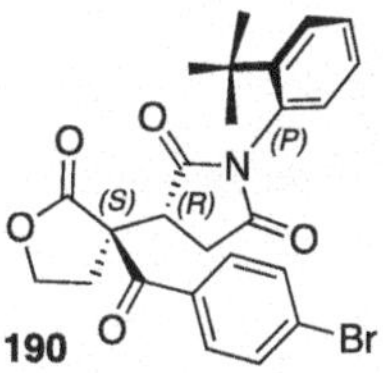

190

The title compound was obtained following the general procedure to furnish the crude with a d.r. of 5:1. The crude mixture was purified by flash column chromatography (hexane:ethyl acetate = 50:50) to give **190** in 63% yield and 40% ee. The ee was determined by HPLC analysis on a Daicel Chiralpak AD-H column: hexane/i-PrOH 80/20, flow rate 0.75 mL/min, λ = 214 nm: **190** τ_{major} = 23.18 min; τ_{minor} = 16.37 min. HRMS-ESI (+): calculated for $C_{25}H_{25}BrNO_5$ 498.0911, found 498.0905 $[M+H]^+$. ^1H NMR (400 MHz, $CDCl_3$): δ (ppm): 7.73–7.57 (5H, m), 7.44–7.37 (3H, m), 4.72–4.56 (2 H, m), 3.56–3.50 (1H, m), 3.26 (1H, q, J_1 = 5.1 Hz, J_2 = 10.1 Hz), 3.14 (1H, dd, J_1 = 5.3 Hz, J_2 = 18.7 Hz), 3.02–2.92 (2H, m), 1.32 (9H, s). ^{13}C NMR (100 MHz, $CDCl_3$): δ (ppm): 209.9 193.9, 177.4, 175.8, 174.5, 148.3, 147.8, 132.4, 132.1, 130.4, 129.8, 128.4, 127.7, 66.4, 60.0, 41.2, 35.5, 32.4, 32.1, 31.6.

***(P)*-*(R)*-3-((*S*)-3-(4-bromobenzoyl)-2-oxotetrahydrofuran-3-yl)-1-(2,5-di-tert-butylphenyl)pyrrolidine-2,5-dione** (product **191**, Table 3.5—entry 9)

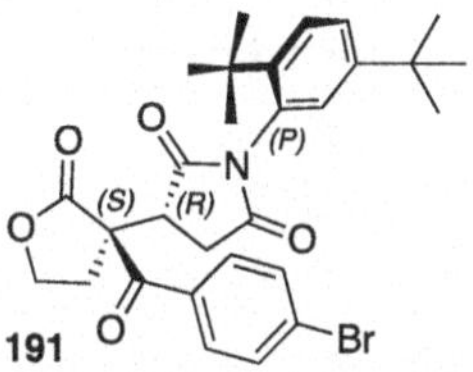

191

The title compound was obtained following the general procedure to furnish the crude with a d.r. of 4:1. The crude mixture was purified by flash column chromatography (hexane:ethyl acetate = 60:40) to give **191** in 56% yield and 30% ee. The ee was determined by HPLC analysis on a Daicel Chiralpak AD-H column: hexane/i-PrOH 80/20, flow rate 0.75 mL/min, λ = 214 nm: **191** τ_{major} = 8.79 min; τ_{minor} = 10.33 min. HRMS-ESI (+): calculated for $C_{29}H_{33}BrNO_5$ 554.1537, found 554.1533 $[M+H]^+$. ^1H NMR (400 MHz, $CDCl_3$): δ (ppm): 7.70–7.62 (4H, m), 7.48 (1H, d, J = 8.5 Hz), 7.41 (1H, dd, J_1 = 2.1 Hz, J_2 = 8.5 Hz), 7.36 (1H, d, J = 2.1 Hz), 4.63–4.56 (2H, m), 3.56–3.50 (1H, m), 3.27 (1H, q, J_1 = 5.8 Hz, J_2 = 10.0 Hz), 3.15 (1H, dd, J_1 = 5.2 Hz, J_2 = 18.7 Hz), 3.03–2.91 (2H, m), 1.35 (9H, s), 1.30 (9H, s). ^{13}C NMR (100 MHz, $CDCl_3$): δ (ppm): 191.7, 177.4, 176.0, 174.5, 150.5, 144.5, 132.4, 132.3, 132.1, 131.0, 130.4, 127.9, 127.2, 126.8, 67.9, 60.0, 48.1, 41.2, 35.1, 34.3, 32.4, 31.6, 32.1, 25.7.

(*M*)-tert-butyl (*R*)-2-((*S*)-1-(2-(tert-butyl)phenyl)-2,5-dioxopyrrolidin-3-yl)-2-cyano-2-phenylacetate (product **192**, Table 3.6—entry 1)

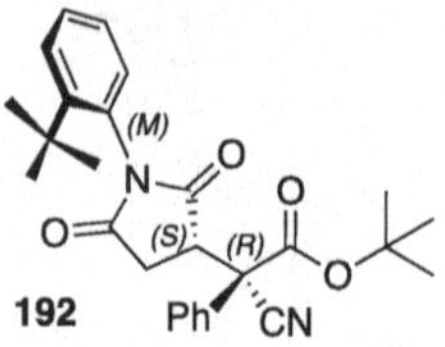

192

The title compound was obtained following the general procedure to furnish the crude product as a single diastereoisomer. The crude mixture was purified by flash column chromatography (hexane:ethyl acetate = 75:25) to give **192** in 82% yield and 92% ee. The ee was determined by HPLC analysis on a Daicel Chiralpak AD-H column: hexane/*i*-PrOH 80/20, flow rate 0.75 mL/min, λ = 214 nm: **192** τ_{major} = 12.13 min; τ_{minor} = 15.02 min. $[\alpha]_D^{20}$ −33.0 (*c* 1.00, $CHCl_3$). HRMS-ESI (+): calculated for $C_{27}H_{31}N_2O_4$ 447.2278, found 447.2277 $[M+H]^+$. ^1H NMR (400 MHz, $CDCl_3$): δ (ppm): 7.63–7.72 (m, 2H); 7.56 (dd, J_1 = 1.4 Hz, J_2 = 7.6 Hz, 1H); 7.44–7.52 (m, 3H); 7.40 (td, J_1 = 7.6 Hz, J_2 = 1.6 Hz, 1H); 7.31 (td, J_1 = 7.6 Hz, J_2 = 1.6 Hz, 1H); 7.02 (dd, J_1 = 1.6 Hz, J_2 = 7.7 Hz, 1H); 4.20–4.42 (m, 3H); 2.86 (dd, J_1 = 9.7 Hz, J_2 = 18.8 Hz, 1H); 2.54 (dd, J_1 = 6.0 Hz, J_2 = 18.8 Hz, 1H); 1.31 (s, 9H); 1.30 (t, J_1 = 7.1, 3H). ^{13}C NMR (75 MHz, $CDCl_3$): δ (ppm): 175.1 (C), 174.1 (C), 166.0 (C), 147.7 (C), 131.3 (C), 131.0 (CH), 130.1 (CH), 129.9 (CH), 129.8 (C), 129.7 $(CH)_2$, 128.6 (CH), 127.8 (CH), 126.6 $(CH)_2$, 116.1 (C), 64.1 (CH_2), 55.5 (C), 46.9 (CH), 35.6 (C), 31.9 (CH_2), 31.8 $(CH_3)_3$, 13.8 (CH_3).

(*M*)-tert-butyl (*R*)-2-((*S*)-1-(4-bromo-2-(tert-butyl)phenyl)-2,5-dioxopyrrolidin-3-yl)-2-cyano-2-phenylacetate (product **193**, Table 3.6—entry 2)

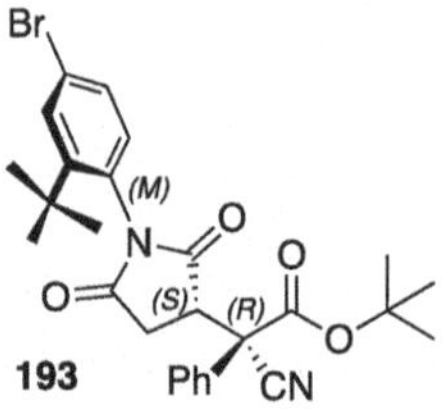

193

The title compound was obtained following the general procedure to furnish the crude product as a single diastereoisomer. The crude mixture was purified by flash column chromatography (hexane:ethyl acetate = 80:20) to give **193** in 67% yield and 81% ee. The ee was determined by HPLC analysis on a Daicel Chiralpak AD-H column: hexane/*i*-PrOH 80/20, flow rate 1 mL/min, λ = 214 nm: **193** τ_{major} = 8.70 min; τ_{minor} = 7.27 min. $[\alpha]_D^{20}$ −16.9 (*c* 1.00, $CHCl_3$). HRMS-ESI (+): calculated for $C_{27}H_{30}BrN_2O_4$ 525.1383, found 525.1368 $[M+H]^+$. ^1H NMR (400 MHz, $CDCl_3$): δ (ppm): 7.68 (*d*, 1H, *J* = 2.2 Hz); 7.63 (*m*, 2H); 7.50-7.41 (*m*, 4H); 6.91 (*d*, 1H, *J* = 8.3 Hz); 4.30 (*dd*, 1H, J_1 = 9.6 Hz, J_2 = 5.7 Hz); 2.84 (*dd*, 1H, J_1 = 19.2 Hz, J_2 = 9.6 Hz); 2.51 (*dd*, 1H, J_1 = 19.2 Hz, J_2 = 5.7 Hz); 1.47 (*s*, 9H); 1.29 (*s*, 9H). ^{13}C NMR (100 MHz, $CDCl_3$): δ (ppm): 175.1, 174.1, 164.4, 150.2, 132.8, 132.1,

131.8, 131.1, 129.9, 129.7, 129.2, 126.5, 124.5, 116.6, 85.9, 56.5, 46.7, 35.9, 32.1, 31.6, 27.7.

(*M*)-tert-butyl (*R*)-2-((*S*)-1-(2-(tert-butyl)-4-chlorophenyl)-2,5-dioxopyrrolidin-3-yl)-2-cyano-2-phenylacetate (product **194**, Table 3.6—entry 3)

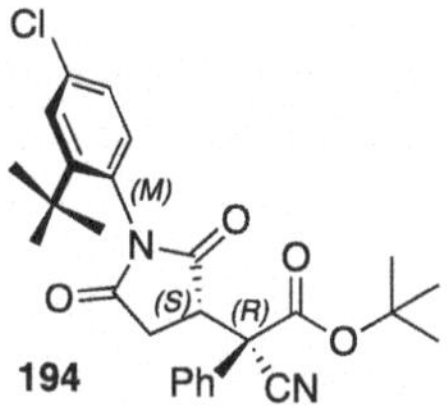

194

The title compound was obtained following the general procedure to furnish the crude product as a single diastereoisomer. The crude mixture was purified by flash column chromatography (hexane:ethyl acetate = 80:20) to give **194** in 71% yield and 80% ee. The ee was determined by HPLC analysis on a Daicel Chiralpak AD-H column: hexane/*i*-PrOH 80/20, flow rate 1 mL/min, λ = 214 nm: **194** τ_{major} = 8.61 min; τ_{minor} = 6.89 min. $[\alpha]_D^{20}$ −38.0 (*c* 1.00, $CHCl_3$). HRMS-ESI (+): calculated for $C_{27}H_{30}ClN_2O_4$ 481.1889, found 481.1891 $[M+H]^+$. ^{1}H NMR (400 MHz, $CDCl_3$): δ (ppm): 7.63 (*m*, 2H); 7.53 (*d*, 1H, J = 2.3 Hz); 7.50–7.44 (*m*, 3H); 7.29 (*dd*, 1H, J_1 = 8.4 Hz, J_2 = 2.3 Hz); 6.98 (*d*, 1H, J = 8.4 Hz); 4.30 (*dd*, 1H, J_1 = 10.0 Hz, J_2 = 5.7 Hz); 2.84 (*dd*, 1H, J_1 = 19.1 Hz, J_2 = 10.0 Hz); 2.51 (*dd*, 1H, J_1 = 19.2 Hz, J_2 = 5.7 Hz); 1.47 (*s*, 9H); 1.29 (*s*, 9H). ^{13}C NMR (100 MHz, $CDCl_3$): δ (ppm): 175.1, 174.2, 164.5, 150.0, 136.1, 132.6, 131.8, 129.9, 129.7, 129.1, 128.1, 126.5, 116.6, 85.9, 56.5, 46.7, 35.9, 32.1, 31.6, 27.7.

(*M*)-tert-butyl (*R*)-2-((*S*)-1-(2-(tert-butyl)-5-nitrophenyl)-2,5-dioxopyrrolidin-3-yl)-2-cyano-2-phenylacetate (product **195**, Table 3.6—entry 4)

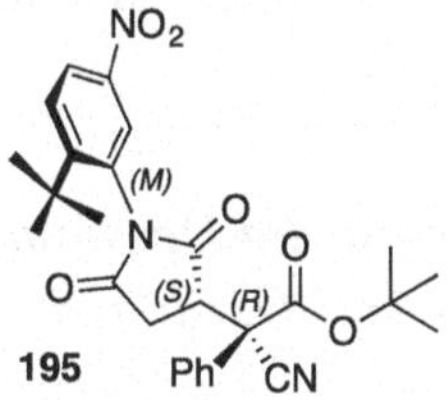

195

The title compound was obtained following the general procedure to furnish the crude product as a 13:1 mixture of diastereoisomers. The crude mixture was purified by flash column chromatography (hexane:ethyl acetate = 80:20) to give **195** in 81% yield and 45% ee. The ee was determined by HPLC analysis on a Daicel Chiralpak AD-H column: hexane/*i*-PrOH 80/20, flow rate 1 mL/min, λ = 214 nm: **195** τ_{major} = 12.20 min; τ_{minor} = 10.24 min. $[\alpha]_D^{20}$ −9.6 (*c* 1.00, $CHCl_3$). HRMS-ESI

(+): calculated for $C_{27}H_{29}N_3NaO_6$ 514.1949, found 514.1951 $[M+Na]^+$. ^{1}H NMR (400 MHz, $CDCl_3$): δ (ppm): 8.23 (*dd*, 1H, $J_1 = 8.9$ Hz, $J_2 = 2.5$ Hz); 7.92 (*d*, 1H, $J = 2.5$ Hz); 7.76 (*d*, 1H, $J = 8.9$ Hz); 7.63 (*m*, 2H); 7.50–7.44 (*m*, 3H); 4.32 (*dd*, 1H, $J_1 = 9.7$ Hz, $J_2 = 5.8$ Hz); 2.93 (*dd*, 1H, $J_1 = 19.0$ Hz, $J_2 = 9.7$ Hz); 2.61 (*dd*, 1H, $J_1 = 19.0$ Hz, $J_2 = 5.8$ Hz); 1.48 (*s*, 9H); 1.34 (*s*, 9H). ^{13}C-NMR (100 MHz, $CDCl_3$): δ (ppm): 174.8, 173.8, 164.5, 156.0, 146.9, 131.6, 131.1, 130.1, 130.0, 129.8, 126.7, 124.6, 116.5, 86.1, 56.3, 46.9, 36.6, 32.3, 31.5, 27.6.

(*M*)-tert-butyl (*R*)-2-((*S*)-1-(4-bromo-2,5-di-tert-butylphenyl)-2,5-dioxopyrrolidin-3-yl)-2-cyano-2-phenylacetate (product **196**, Table 3.6—entry 5)

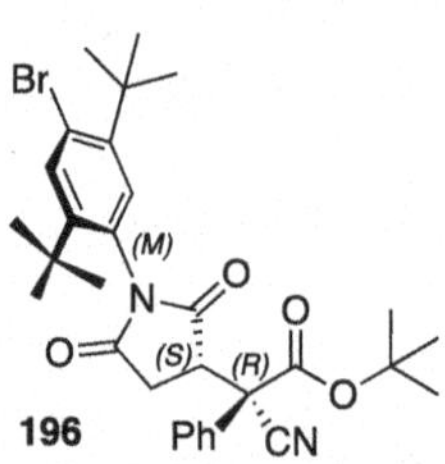

196

The title compound was obtained following the general procedure to furnish the crude product as a single diastereoisomer. The crude mixture was purified by flash column chromatography (hexane:ethyl acetate = 80:20) to give **196** in 50% yield and 50% ee. The ee was determined by HPLC analysis on a Daicel Chiralpak AD-H column: hexane/*i*-PrOH 80/20, flow rate 1 mL/min, $\lambda = 214$ nm: **196** $\tau_{major} = 5.06$ min; $\tau_{minor} = 5.94$ min. $[\alpha]_D^{20}$ −9.1 (*c* 1.00, $CHCl_3$). HRMS-ESI (+): calculated for $C_{31}H_{38}BrN_2O_4$ 514.1949, found 514.1951 $[M+H]^+$. ^{1}H NMR (400 MHz, $CDCl_3$): δ (ppm): 7.72 (*s*, 1H); 7.62 (*m*, 2H); 7.50–7.42 (*m*, 3H); 7.05 (*s*, 1H); 4.29 (*dd*, 1H, $J_1 = 9.8$ Hz, $J_2 = 5.6$ Hz); 2.84 (*dd*, 1H, $J_1 = 18.9$ Hz, $J_2 = 9.8$ Hz); 2.51 (*dd*, 1H, $J_1 = 18.9$ Hz, $J_2 = 5.6$ Hz); 1.47 (*s*, 18H); 1.27 (*s*, 9H). ^{13}C NMR (100 MHz, $CDCl_3$): δ: δ (ppm): 175.2, 174.4, 164.5, 148.5, 147.7, 146.9, 136.0, 131.9, 130.8, 129.9, 128.9, 128.0, 124.6, 116.5, 85.8, 56.7, 46.6, 36.4, 35.2, 32.2, 31.5, 29.5, 27.7.

3.4.5 General Procedure for the Desymmetrization of Maleimides with 3-Aryl Oxindoles

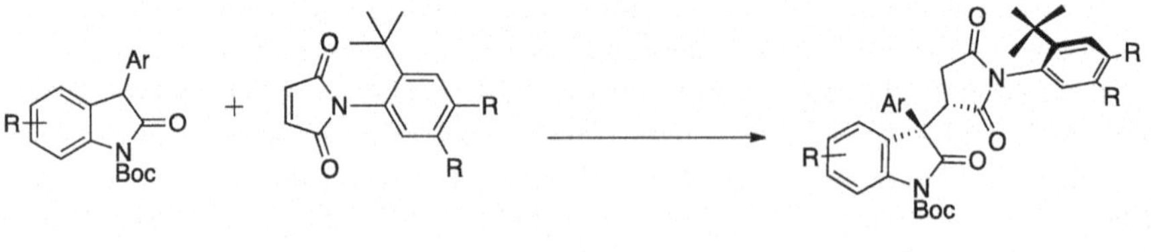

All the reactions were carried out in undistilled solvent and stirring was provided by magnetic Teflon-coated stir bars. In an ordinary vial were placed catalyst **172** (0.02 mmol, 0.1 equiv.), the maleimide (0.2 mmol, 1.0 equiv.) and the oxindole (0.21 mmol, 1.05 equiv.) before adding the solvent (DCM, 0.8 mL; 0.25 M) and after 24 h under magnetic stirring, the crude mixture was flushed through a short plug of silica, using dichloromethane/ethyl acetate 1:1 as the eluent (50 ml). Then solvent was removed in *vacuo* and the diastereomeric ratio (dr) was determined by ^{1}H NMR analysis of the crude mixture. Finally, the desired compound was isolated by flash column chromatography and the enantiomeric excess was determined by means of chiral HPLC analysis.

(*M*)-(*R*)-*tert*-butyl-3-((*S*)-1-(2-(*tert*-butyl)phenyl)-2,5-dioxopyrrolidin-3-yl)-2-oxo-3-phenylindoline-1-carboxylate (product **198**, Table 3.7—entry 1)

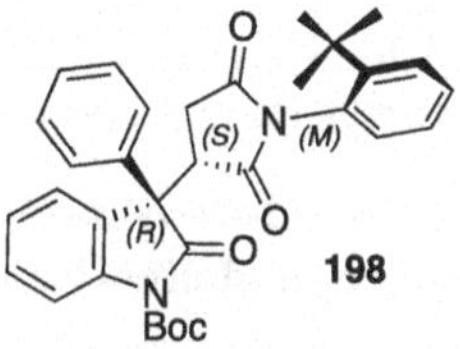

198

The title compound was obtained following the general procedure to furnish the crude product as a single diastereoisomer (dr > 19:1). The crude mixture was purified by flash column chromatography (hexane:diethyl ether = 6:4) to give of **198** in 82% yield 0.164 mmol, and > 99% ee. HPLC analysis on a Daicel Chiralpak AD-H column: hexane/*i*-PrOH 90/10, flow rate 1 mL/min, λ = 254 nm: **198** τ_{major} = 19.0 min; τ_{minor} = 8.3 min. ^{1}H NMR (400 MHz, $CDCl_3$): δ (ppm): 8.07 (*d*, J = 8.3 Hz, 1H); 7.53 (*m*, 1H,); 7.45 (*dd*, J_1 = 8.2 Hz, J_2 = 1.5 Hz, 1H); 7.34 (*m*, 7H); 7.25 (*m*, 1H,); 7.01 (*ddd*, J_1 = J_2 = 7.6 Hz, J_3 = 1.4 Hz, 1H); 5.71 (*dd*, J_1 = 7.7 Hz, J_2 = 1.4 Hz, 1H); 4.44 (*dd*, J_1 = 10.3 Hz, J_2 = 4.0 Hz, 1H); 3.11 (*dd*, J_1 = 19.3 Hz, J_2 = 10.3 Hz, 1H); 2.81 (*dd*, J_1 = 19.3 Hz, J_2 = 4.0 Hz, 1H); 1.57 (*s*, 9H); 1.23 (*s*, 9H). ^{13}C NMR (100 MHz, $CDCl_3$): δ (ppm): 175.8, 175.7, 174.3, 148.9, 147.7, 141.5, 136.8, 134.9, 130.0, 129.9, 129.7, 129.6, 129.1, 128.5, 127.6, 127.3, 126.4, 124.4, 123.8, 116.6, 84.6, 56.9, 47.8, 35.5, 32.8, 31.6, 28.0. $[\alpha]_D^{20}$ +193.5 (*c* 1.00, $CHCl_3$). HRMS-ESI-ORBITRAP (+): calculated for $C_{33}H_{34}N_2NaO_5$ 561.236, found 561.2359 $(M+Na)^+$. Calculated for $C_{33}H_{34}KN_2O_5$ 577.2099, found 577.2096 $(M+K)^+$.

(*M*)-(*R*)-*tert*-butyl-3-((*S*)-1-(2-(*tert*-butyl)phenyl)-2,5-dioxopyrrolidin-3-yl)-5-fluoro-2-oxo-3-phenylindoline-1-carboxylate (product **199**, Table 3.7—entry 2)

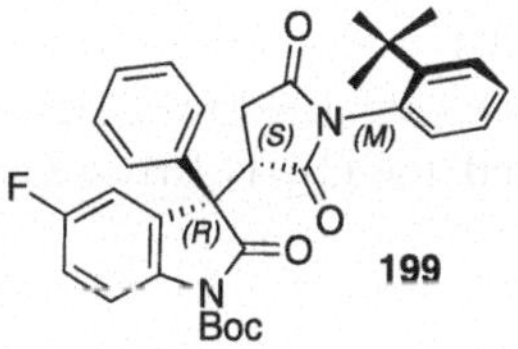

199

The title compound was obtained following the general procedure to furnish the crude product as a 10:1 mixture of diastereoisomers. The crude mixture was purified by flash column chromatography (dichloromethane:hexane = 9:1) to give **199** in 90% yield, 0.18 mmol and 98% ee. HPLC analysis on a Daicel Chiralpak AD-H column: hexane/*i*-PrOH 70/30, flow rate 0.8 mL/min, $\lambda = 254$ nm: **199** $\tau_{major} = 8.3$ min; $\tau_{minor} = 5.8$ min. ^{1}H NMR (400 MHz, $CDCl_3$): δ (ppm): 8.08 (*dd*, $J_1 = 9.2$ Hz, $J_2 = 4.7$ Hz, 1H); 7.48 (*dd*, $J_1 = 8.1$ Hz, $J_2 = 1.3$ Hz, 1H); 7.42–7.19 (*m*, 7H); 7.09 (*m*, 2H,); 5.91 (*dd*, $J_1 = 7.8$ Hz, $J_2 = 1.4$ Hz, 1H); 4.46 (*dd*, $J_1 = 10.2$ Hz, $J_2 = 4.3$ Hz, 1H); 3.13 (*dd*, $J_1 = 19.5$ Hz, $J_2 = 10.6$ Hz, 1H); 2.74 (*dd*, $J_1 = 19.5$ Hz, $J_2 = 4.3$ Hz, 1H); 1.57 (*s*, 9H); 1.24 (*s*, 9H). ^{13}C NMR (100 MHz, $CDCl_3$): δ (ppm): 175.8, 175.2, 174.1, 160.9, 158.5, 148.9, 147.8, 137.6, 137.5, 136.3, 129.9 (*d*, $J = 1.5$ Hz), 129.5, 129.3, 128.8 (*d*, $J = 5.4$ Hz), 128.2 (*d*, $J = 7.4$ Hz), 127.5, 118.0 (*d*, $J = 8.6$ Hz), 116.5 (*d*, $J = 22.5$ Hz), 111.2 (*d*, $J = 24.3$ Hz), 84.9, 57.0, 47.7, 35.6, 32.7, 31.6, 28.0. ^{19}F NMR (376 MHz, $CDCl_3$): δ (ppm): −116.6 (1F). $[\alpha]_D^{20}$ +123.5 (*c* 2.00, $CHCl_3$). HRMS-ESI-ORBITRAP (+): calculated for $C_{33}H_{33}FN_2NaO_5$ 579.2266, found 579.2263 $(M+Na)^+$. Calculated for $C_{33}H_{33}FN_2KO_5$ 595.2005, found 595.1988 $(M+K)^+$.

(*M*)-(*R*)-*tert*-butyl-6-bromo-3-((*S*)-1-(2-(*tert*-butyl)phenyl)-2,5-dioxopyrrolidin-3-yl)-2-oxo-3-phenylindoline-1-carboxylate (product **200**, Table 3.7—entry 3)

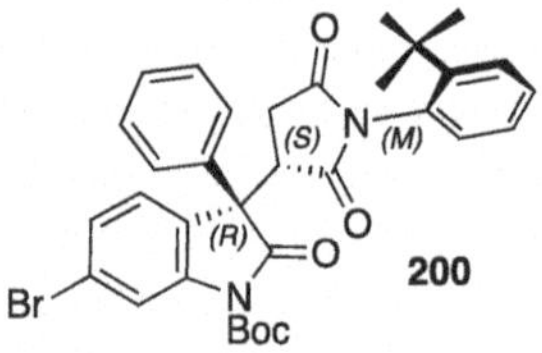

200

The title compound was obtained following the general procedure to furnish the crude product as a single diastereoisomer (dr 19:1). The crude mixture was purified by flash column chromatography (hexane:diethyl ether = 6:4) to give **200** in 81% yield, 0.162 mmol and 98% ee. HPLC analysis on a Daicel Chiralpak AD-H column: hexane/*i*-PrOH 90/10, flow rate 1 mL/min, $\lambda = 254$ nm: **200** $\tau_{major} = 13.5$ min; $\tau_{minor} = 8.5$ min. ^{1}H NMR (300 MHz, $CDCl_3$): δ (ppm): 8.32 (*d*, $J = 1.9$ Hz, 1H); 7.48 (*m*, 2H); 7.40–7.24 (*m*, 7H); 7.22 (*d*, $J = 8.1$ Hz, 1H 2H,); 7.11 (*ddd*, $J_1 = J_2 = 7.6$ Hz, $J_3 = 1.4$ Hz, 1H); 5.78 (*dd*, $J_1 = 7.8$ Hz, $J_2 = 1.5$ Hz, 1H); 4.42 (*dd*, $J_1 = 10.2$ Hz, $J_2 = 4.0$ Hz, 1H); 3.11 (*dd*, $J_1 = 19.3$ Hz, $J_2 = 10.2$ Hz, 1H); 2.75 (*dd*, $J_1 = 19.3$ Hz, $J_2 = 4.0$ Hz, 1H); 1.57 (*s*, 9H); 1.24 (*s*, 9H). ^{13}C NMR (100 MHz, $CDCl_3$): δ (ppm): 175.7, 175.4, 173.8, 148.7, 147.7, 142.5, 136.2, 134.9, 131.3, 129.8, 129.5, 129.3, 128.7, 128.6, 127.5, 127.4, 125.3, 124.9, 120.1, 85.1, 56.7, 47.7, 35.5, 32.7, 31.6, 27.9. $[\alpha]_D^{20}$ +313.1 (*c* 2.00, $CHCl_3$).

HRMS-ESI-ORBITRAP (+): calculated for $C_{33}H_{33}BrN_2NaO_5$ 639.1465, found 639.1462 $(M+Na)^+$. Calculated for $C_{33}H_{33}BrN_2KO_5$ 655.1204, found 655.1163 $(M+K)^+$.

(*M*)-(*R*)-*tert*-butyl-3-((*S*)-1-(2-(*tert*-butyl)phenyl)-2,5-dioxopyrrolidin-3-yl)-7-fluoro-2-oxo-3-phenylindoline-1-carboxylate (product **201**, Table 3.7—entry 4)

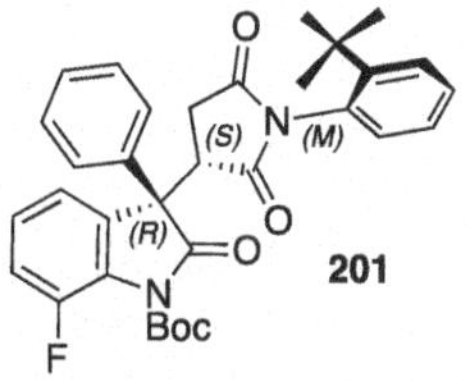

201

The title compound was obtained following the general procedure to furnish the crude product as a single diastereoisomer (dr 19:1). The crude mixture was purified by flash column chromatography (hexane:diethyl ether = 1:1) to give **201** in 98% yield, 0.196 mmol and 98% ee. HPLC analysis on a Daicel Chiralpak AD-H column: hexane/*i*-PrOH 70/30, flow rate 0.8 mL/min, $\lambda = 254$ nm: **201** $\tau_{major} = 21.3$ min; $\tau_{minor} = 7.3$ min. ^{1}H NMR (400 MHz, $CDCl_3$): δ (ppm): 7.46 (*dd*, $J_1 = 8.3$ Hz, $J_2 = 1.2$ Hz, 1H); 7.42–7.24 (*m*, 8H); 7.17 (*dd*, $J_1 = 6.6$ Hz, $J_2 = 1.9$ Hz, 1H); 7.03 (*ddd*, $J_1 = J_2 = 7.7$ Hz, $J_3 = 1.3$ Hz, 1H); 5.76 (*dd*, $J_1 = 7.7$ Hz, $J_2 = 1.1$ Hz, 1H); 4.43 (*dd*, $J_1 = 10.2$ Hz, $J_2 = 4.0$ Hz, 1H); 3.11 (*dd*, $J_1 = 19.2$ Hz, $J_2 = 10.2$ Hz, 1H); 2.79 (*dd*, $J_1 = 19.2$ Hz, $J_2 = 4.0$ Hz, 1H); 1.54 (*s*, 9H); 1.24 (*s*, 9H). ^{13}C NMR (100 MHz, $CDCl_3$): δ (ppm): 175.5, 175.4, 173.9, 150.6, 148.1, 147.7, 147.0, 136.2, 129.9 (*d*, $J = 2.2$ Hz), 129.8, 129.7, 129.6, 129.2, 128.6 (*d*, $J = 12.5$ Hz), 128.6 (*d*, $J = 3.7$ Hz), 127.4, 127.2, 125.5 (*d*, $J = 7.0$ Hz), 119.7 (*d*, $J = 3.7$ Hz), 118.4, (*d*, $J = 20.5$ Hz), 85.1, 57.5, 47.7, 35.5, 32.7, 32.6, 27.6.

^{19}F NMR (376 MHz, $CDCl_3$): δ (ppm): −115.5 (1F). $[\alpha]_D^{20}$ +173.0 (*c* 2.00, $CHCl_3$). HRMS-ESI-ORBITRAP (+): calculated for $C_{33}H_{33}FN_2NaO_5$ 579.2266, found 579.2266 $(M+Na)^+$. Calculated for $C_{33}H_{33}FN_2KO_5$ 595.2005, found 595.1996 $(M+K)^+$.

(*M*)-(*R*)-*tert*-butyl-3-((*S*)-1-(2-(*tert*-butyl)phenyl)-2,5-dioxopyrrolidin-3-yl)-5-methoxy-2-oxo-3-phenylindoline-1-carboxylate (product **202**, Table 3.7—entry 5)

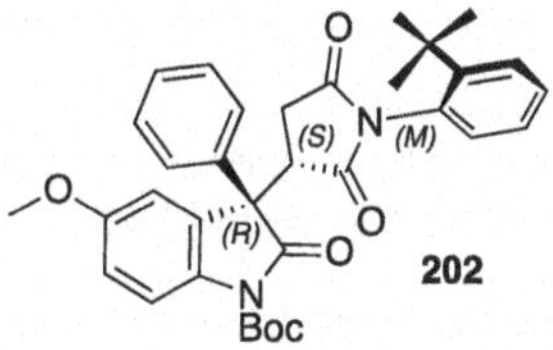

202

The title compound was obtained following the general procedure to furnish the crude product as a single diastereoisomer (dr > 19:1). The crude mixture was purified by flash column chromatography (hexane:ethyl acetate = 75:25) to give **202** in 77% yield, 0.154 mmol and 98% ee. HPLC analysis on a Daicel Chiralpak AD-H column:

hexane/*i*-PrOH 80/20, flow rate 1 mL/min, λ = 254 nm: **202** τ_{major} = 12.5 min; τ_{minor} = 6.4 min. ^{1}H NMR (400 MHz, $CDCl_3$): δ (ppm): 8.00 (*d*, *J* = 9.1 Hz, 1H); 7.46 (*dd*, J_1 = 8.2 Hz, J_2 = 1.3 Hz, 1H); 7.41–7.31 (*m*, 5H); 7.27 (*m*, 1H,); 7.04 (*m*, 2H,); 6.88 (*d*, *J* = 2.6 Hz, 1H); 5.76 (*dd*, J_1 = 7.9 Hz, J_2 = 1.6 Hz, 1H); 4.44 (*dd*, J_1 = 10.1 Hz, J_2 = 3.6 Hz, 1H); 3.80 (*s*, 3H); 3.12 (*dd*, J_1 = 19.3 Hz, J_2 = 10.1 Hz, 1H); 2.79 (*dd*, J_1 = 19.3 Hz, J_2 = 3.6 Hz, 1H); 1.57 (*s*, 9H); 1.24 (*s*, 9H). ^{13}C NMR (100 MHz, $CDCl_3$): δ (ppm): 175.8, 175.7, 174.3, 156.8, 149.0, 147.8, 136.8, 134.7, 129.7, 129.6, 129.6, 129.1, 128.5, 128.4, 127.6, 127.3, 127.2, 117.5, 115.0, 109.7, 84.3, 57.3, 55.8, 47.7, 35.5, 32.9, 31.5, 28.0. $[\alpha]_D^{20}$ +105.4 (*c* 2.00, $CHCl_3$). HRMS-ESI-ORBITRAP (+): calculated for $C_{34}H_{36}N_2NaO_6$ 591.2466, found 591.2455 $(M+Na)^+$. Calculated for $C_{34}H_{36}N_2KO_6$ 607.2205, found 607.2182 $(M+K)^+$.

***(M)*-*(R)*-*tert*-butyl-3-((*S*)-1-(2-(*tert*-butyl)phenyl)-2,5-dioxopyrrolidin-3-yl)-5-methyl-2-oxo-3-phenylindoline-1-carboxylate** (product **203**, Table 3.7—entry 6)

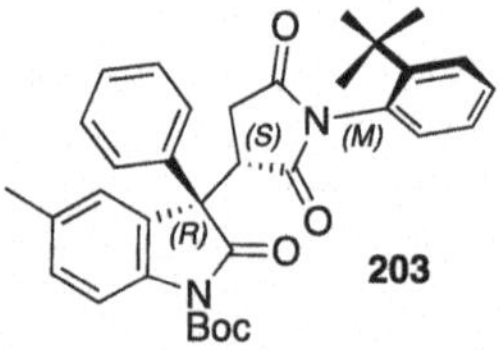

203

The title compound was obtained following the general procedure to furnish the crude product as a single diastereoisomer (dr > 19:1). The crude mixture was purified by flash column chromatography (dichloromethane:hexane = 90/10) to give **203** in 82% yield, 0.164 mmol and 98% ee. HPLC analysis on a Daicel Chiralpak AD-H column: hexane/*i*-PrOH 70/30, flow rate 0.8 mL/min, λ = 254 nm: **203** τ_{major} = 8.5 min; τ_{minor} = 5.4 min. ^{1}H NMR (400 MHz, $CDCl_3$): δ (ppm): 7.94 (*d*, *J* = 8.4 Hz, 1H); 7.46 (*dd*, J_1 = 8.2 Hz, J_2 = 1.3 Hz, 1H); 7.39–7.30 (*m*, 6H); 7.26 (*ddd*, J_1 = 8.8, J_2 = 7.4 Hz, J_3 = 1.5 Hz, 1H); 7.15 (*bs*, 1H,); 7.03 (*ddd*, J_1 = J_2 = 7.0 Hz, J_3 = 1.4 Hz, 1H); 5.65 (*dd*, J_1 = 7.8 Hz, J_2 = 1.5 Hz, 1H); 4.42 (*dd*, J_1 = 10.1 Hz, J_2 = 3.6 Hz, 1H); 3.10 (*dd*, J_1 = 19.2 Hz, J_2 = 10.1 Hz, 1H); 2.81 (*dd*, J_1 = 19.2 Hz, J_2 = 3.6 Hz, 1H); 2.43 (*s*, 3H); 1.57 (*s*, 9H); 1.24 (*s*, 9H). ^{13}C NMR (100 MHz, $CDCl_3$): δ (ppm): 175.9, 175.8, 174.4, 149.0, 147.8, 139.1, 136.9, 134.2, 130.4, 129.8, 129.7, 129.6, 129.1, 128.5, 128.4, 127.7, 127.2, 126.3, 124.3, 116.4, 84.4, 57.1, 47.8, 35.5, 32.9, 31.6, 28.0, 21.1. $[\alpha]_D^{20}$ +84.7 (*c* 1.00, $CHCl_3$). HRMS-ESI-ORBITRAP (+): calculated for $C_{34}H_{36}N_2NaO_5$ 575.2516, found 575.2517 $(M+Na)^+$. Calculated for $C_{34}H_{36}N_2KO_5$ 591.2256, found 591.2278 $(M+K)^+$.

***(M)*-*(R)*-*tert*-butyl-3-((*S*)-1-(2-(*tert*-butyl)phenyl)-2,5-dioxopyrrolidin-3-yl)-5,7-dimethyl-2-oxo-3-phenylindoline-1-carboxylate** (product **204**, Table 3.7—entry 7)

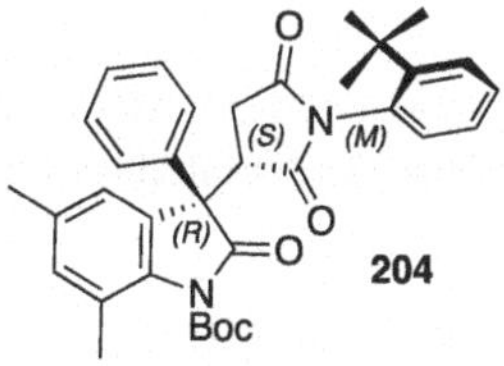

204

The title compound was obtained following the general procedure to furnish the crude product as a 10:1 mixture of diastereoisomers. The crude mixture was purified by flash column chromatography (dichloromethane:hexane = 90/10) to give **204** in 50% yield, 0.100 mmol and 93% ee. HPLC analysis on a Daicel Chiralpak IC column: hexane/*i*-PrOH 80/20, flow rate 1 mL/min, $\lambda = 254$ nm: **204** $\tau_{major} = 9.9$ min; $\tau_{minor} = 25.1$ min. ^{1}H NMR (400 MHz, $CDCl_3$): δ (ppm): 7.44 (*dd*, $J_1 = 8.1$ Hz, $J_2 = 1.4$ Hz, 1H); 7.40–7.30 (*m*, 5H); 7.25 (*ddd*, $J_1 = 8.9$, $J_2 = 7.4$ Hz, $J_3 = 1.5$ Hz, 1H); 7.14 (*bs*, 1H,); 7.02 (*ddd*, $J_1 = J_2 = 7.6$ Hz, $J_3 = 1.4$ Hz, 1H); 6.98 (*bs*, 1H,); 5.66 (*dd*, $J_1 = 7.8$ Hz, $J_2 = 1.4$ Hz, 1H); 4.38 (*dd*, $J_1 = 10.2$ Hz, $J_2 = 3.6$ Hz, 1H); 3.08 (*dd*, $J_1 = 19.4$ Hz, $J_2 = 10.2$ Hz, 1H); 2.79 (*dd*, $J_1 = 19.4$ Hz, $J_2 = 3.6$ Hz, 1H); 2.38 (*s*, 3H); 2.23 (*s*, 3H); 1.54 (*s*, 9H); 1.23 (*s*, 9H). ^{13}C NMR (100 MHz, $CDCl_3$): δ (ppm): 176.0, 175.6, 175.4, 148.8, 147.8, 137.6, 137.2, 134.2, 133.4, 129.9, 129.8, 129.6, 129.0, 128.5, 128.3, 127.7, 127.6, 127.2, 125.1, 122.0, 84.6, 57.5, 47.6, 35.5, 32.9, 31.6, 27.7, 21.0, 20.0. $[\alpha]_D^{20}$ +100.4 (*c* 1.00, $CHCl_3$). HRMS-ESI-ORBITRAP (+): calculated for $C_{35}H_{38}N_2NaO_5$ 589.2673, found 589.2675 $(M+Na)^+$. Calculated for $C_{35}H_{38}N_2KO_5$ 605.2412, found 605.2397 $(M+K)^+$.

(*M*)-(*R*)-*tert*-butyl-3-((*S*)-1-(2-(*tert*-butyl)phenyl)-2,5-dioxopyrrolidin-3-yl)-2-oxo-3-(p-tolyl)indoline-1-carboxylate (product **205**, Table 3.7—entry 8)

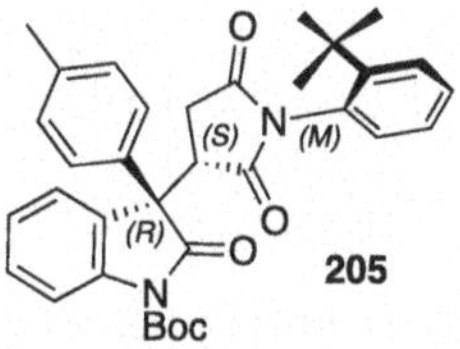

205

The title compound was obtained following the general procedure to furnish the crude product as a 10:1 mixture of diastereoisomers. The crude mixture was purified by flash column chromatography (hexane:ethyl acetate = 80/20) to give **205** in 77% yield, 0.154 mmol and 98% ee. HPLC analysis on a Daicel Chiralpak IC column: hexane/*i*-PrOH 80/20, flow rate 1 mL/min, $\lambda = 254$ nm: **205** $\tau_{major} = 23.9$ min; $\tau_{minor} = 11.9$ min. ^{1}H NMR (400 MHz, $CDCl_3$): δ (ppm): 8.07 (*d*, $J = 8.2$ Hz, 1H); 7.51 (*m*, 1H); 7.45 (*dd*, $J_1 = 8.2$ Hz, $J_2 = 1.4$ Hz, 1H); 7.37-7.30 (*m*, 2H); 7.30–7.11 (*m*, 5H); 7.01 (*ddd*, $J_1 = J_2 = 7.6$ Hz, $J_3 = 1.4$ Hz, 1H); 5.70 (*dd*, $J_1 = 7.9$ Hz, $J_2 = 1.4$ Hz, 1H); 4.42 (*dd*, $J_1 = 10.2$ Hz, $J_2 = 4.0$ Hz, 1H); 3.11 (*dd*, $J_1 = 19.3$ Hz, $J_2 = 10.2$ Hz, 1H); 2.82 (*dd*, $J_1 = 19.3$ Hz, $J_2 = 4.0$ Hz, 1H); 2.33 (*s*, 3H); 1.57 (*s*, 9H); 1.23 (*s*, 9H). ^{13}C NMR (100 MHz, $CDCl_3$): δ (ppm): 175.9, 175.8, 174.4, 149.0, 147.7, 141.5, 138.4, 133.8, 129.9, 129.8, 129.7, 129.6, 129.6, 128.4, 127.5,

127.2, 126.5, 124.4, 123.7, 116.5, 84.5, 56.7, 47.7, 35.5, 32.8, 31.5, 28.0, 20.9. $[\alpha]_D^{20}$ +128.8 (*c* 1.00, $CHCl_3$). HRMS-ESI-ORBITRAP (+): calculated for $C_{34}H_{36}N_2NaO_5$ 575.2516, found 575.2511 $(M+Na)^+$. Calculated for $C_{34}H_{36}N_2KO_5$ 591.2256, found 591.225 $(M+K)^+$.

(*M*)-(*R*)-*tert*-butyl-3-((*S*)-1-(2-(*tert*-butyl)phenyl)-2,5-dioxopyrrolidin-3-yl)-2-oxo-3-(m-tolyl)indoline-1-carboxylate (product **206**, Table 3.7—entry 9)

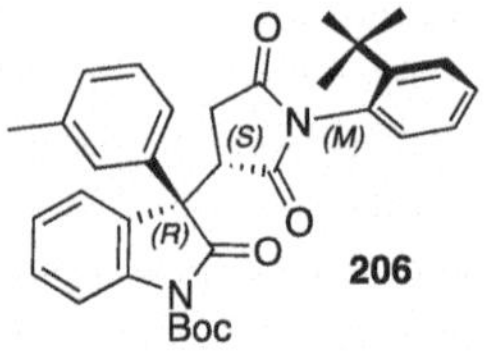

206

The title compound was obtained following the general procedure to furnish the crude product as a single diastereoisomer (dr 19:1). The crude mixture was purified by flash column chromatography (hexane:ethyl acetate = 80/20) to give **206** in 79% yield, 0.158 mmol and 98% ee. HPLC analysis on a Daicel Chiralpak AD-H column: hexane/*i*-PrOH 80/20, flow rate 1 mL/min, $\lambda = 254$ nm: **206** $\tau_{major} = 7.2$ min; $\tau_{minor} = 4.4$ min. ^1H NMR (400 MHz, $CDCl_3$): δ (ppm): 8.07 (*d*, $J = 8.2$ Hz, 1H); 7.52 (*m*, 1H); 7.45 (*dd*, $J_1 = 8.2$ Hz, $J_2 = 1.3$ Hz, 1H); 7.38–7.29 (*m*, 2H); 7.29–7.20 (*m*, 2H); 7.17–7.08 (*m*, 3H); 7.01 (*ddd*, $J_1 = J_2 = 7.6$ Hz, $J_3 = 1.4$ Hz, 1H); 5.70 (*dd*, $J_1 = 7.8$ Hz, $J_2 = 1.5$ Hz, 1H); 4.43 (*dd*, $J_1 = 10.2$ Hz, $J_2 = 3.9$ Hz, 1H); 3.12 (*dd*, $J_1 = 19.2$ Hz, $J_2 = 10.2$ Hz, 1H); 2.82 (*dd*, $J_1 = 19.2$ Hz, $J_2 = 3.9$ Hz, 1H); 2.33 (*s*, 3H); 1.58 (*s*, 9H); 1.23 (*s*, 9H). ^{13}C NMR (100 MHz, $CDCl_3$): δ (ppm): 175.9, 175.8, 174.3, 149.0, 147.7, 141.5, 138.9, 136.7, 129.9, 129.9, 129.6, 129.6, 129.3, 128.9, 128.5, 128.2, 127.2, 126.5, 124.7, 124.4, 123.8, 116.5, 84.5, 56.9, 47.8, 35.5, 32.8, 31.5, 28.0, 21.6. $[\alpha]_D^{20}$ +131.1 (*c* 2.00, $CHCl_3$). HRMS-ESI-ORBITRAP (+): calculated for $C_{34}H_{36}N_2NaO_5$ 575.2516, found 575.253 $(M+Na)^+$. Calculated for $C_{34}H_{36}N_2KO_5$ 591.2256, found 591.2282 $(M+K)^+$.

(*M*)-(*R*)-*tert*-butyl-3-((*S*)-1-(2-(*tert*-butyl)phenyl)-2,5-dioxopyrrolidin-3-yl)-3-(4-methoxyphenyl)-2-oxoindoline-1-carboxylate (product **207**, Table 3.7—entry 10)

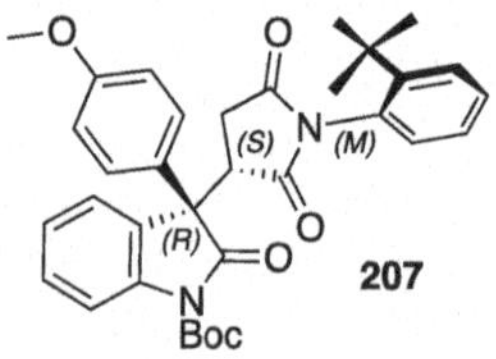

207

The title compound was obtained following the general procedure to furnish the crude product as a single diastereoisomer (dr > 19:1). The crude mixture was purified by flash column chromatography (hexane:ethyl acetate = 90:10) to give **207** in 82%

yield, 0.164 mmol and 96% ee. HPLC analysis on a Daicel Chiralpak AD-H column: hexane/*i*-PrOH 80/20, flow rate 1 mL/min, λ = 254 nm: **207** τ_{major} = 14.6 min; τ_{minor} = 7.2 min. ^{1}H NMR (400 MHz, $CDCl_3$): δ (ppm): 8.06 (*d*, *J* = 8.3 Hz, 1H); 7.52 (*m*, 1H,); 7.45 (*d*, *J* = 8.2 Hz, 1H); 7.34 (*m*, 2H); 7.25 (*m*, 3H,); 7.01 (*m*, 1H,); 6.88 (*m*, 2H); 5.70 (*dd*, J_1 = 7.9 Hz, J_2 = 1.2 Hz, 1H); 4.39 (*dd*, J_1 = 10.3 Hz, J_2 = 3.8 Hz, 1H); 3.78 (*s*, 3H); 3.10 (*dd*, J_1 = 19.3 Hz, J_2 = 10.3 Hz, 1H); 2.83 (*dd*, J_1 = 19.3 Hz, J_2 = 3.8 Hz, 1H); 1.57 (*s*, 9H); 1.23 (*s*, 9H). ^{13}C NMR (75 MHz, $CDCl_3$): δ (ppm): 175.9, 175.9, 174.7, 159.5, 148.9, 147.6, 141.3, 129.9, 129.8, 129.6, 129.5, 128.8, 128.5, 128.4, 127.1, 126.5, 124.3, 123.7, 116.4, 114.4, 84.5, 56.3, 55.2, 47.7, 35.4, 32.6, 31.4, 27.9. $[\alpha]_D^{20}$ +123.6 (*c* 2.00, $CHCl_3$). HRMS-ESI-ORBITRAP (+): calculated for $C_{34}H_{36}N_2NaO_6$ 591.2466, found 591.2467 $(M+Na)^+$. Calculated for $C_{34}H_{36}N_2KO_6$ 607.2205, found 607.2185 $(M+K)^+$.

(*M*)-(*R*)-*tert*-butyl-3-((*S*)-1-(2-(*tert*-butyl)-4-chlorophenyl)-2,5-dioxopyrrolidin-3-yl)-2-oxo-3-phenylindoline-1-carboxylate (product **208**, Table 3.7—entry 11)

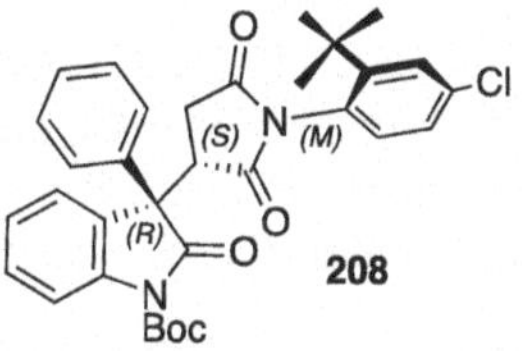

208

The title compound was obtained following the general procedure to furnish the crude product as a single diastereoisomer (dr > 19:1). The crude mixture was purified by flash column chromatography (hexane:ethyl acetate = 75:25) to give **208** in 90% yield, 0.180 mmol and 98% ee. HPLC analysis on a Daicel Chiralpak AD-H column: hexane/*i*-PrOH 80/20, flow rate 1 mL/min, λ = 254 nm: **208** τ_{major} = 7.9 min; τ_{minor} = 5.3 min. ^{1}H NMR (400 MHz, $CDCl_3$): δ (ppm): 8.08 (*d*, *J* = 8.2 Hz, 1H); 7.52 (*m*, 1H,); 7.41 (*d*, *J* = 2.4 Hz, 1H); 7.34 (*m*, 7H); 7.00 (*dd*, J_1 = 8.4 Hz, J_2 = 2.4, 1H); 5.59 (*d*, *J* = 8.5 Hz, 1H); 4.44 (*dd*, J_1 = 10.2 Hz, J_2 = 3.8 Hz, 1H); 3.12 (*dd*, J_1 = 19.5 Hz, J_2 = 10.2 Hz, 1H); 2.81 (*dd*, J_1 = 19.5 Hz, J_2 = 3.8 Hz, 1H); 1.58 (*s*, 9H); 1.22 (*s*, 9H). ^{13}C NMR (100 MHz, $CDCl_3$): δ (ppm): 175.7, 175.5, 174.2, 149.8, 148.9, 141.5, 136.6, 135.7, 131.3, 130.1, 129.1, 128.8, 128.5, 128.3, 127.6, 127.5, 126.3, 124.4, 123.8, 116.6, 84.7, 57.0, 47.8, 35.7, 32.8, 31.3, 28.0. $[\alpha]_D^{20}$ +149.0 (*c* 1.00, $CHCl_3$). HRMS-ESI-ORBITRAP (+): calculated for $C_{33}H_{33}ClN_2NaO_5$ 595.197, found 595.1971 $(M+Na)^+$. Calculated for $C_{33}H_{33}ClN_2KO_5$ 611.171, found 611.1714 $(M+K)^+$.

(*M*)-(*R*)-*tert*-butyl-3-((*S*)-1-(4-bromo-2-(*tert*-butyl)phenyl)-2,5-dioxopyrrolidin-3-yl)-2-oxo-3-phenylindoline-1-carboxylate (product **209**, Table 3.7—entry 12)

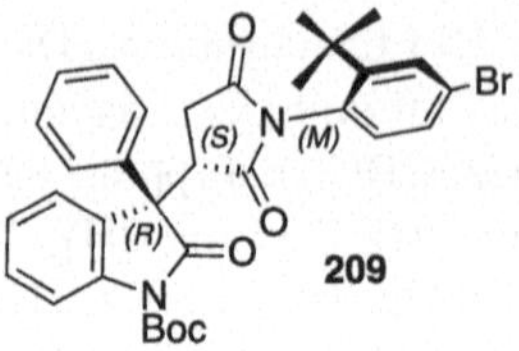

209

The title compound was obtained following the general procedure to furnish the crude product as a 16:1 mixture of diastereoisomers. The crude mixture was purified by flash column chromatography (hexane:ethyl acetate = 80:20) to give **209** in 79% yield, 0.158 mmol and 96% ee. HPLC analysis on a Daicel Chiralpak AD-H column: hexane/*i*-PrOH 80/20, flow rate 0.8 mL/min, λ = 254 nm: **209** τ_{major} = 10.2 min; τ_{minor} = 6.7 min. ^{1}H NMR (400 MHz, $CDCl_3$): δ (ppm): 8.07 (*d*, *J* = 8.2 Hz, 1H); 7.57 (*d*, *J* = 2.2 Hz, 1H); 7.52 (*m*, 1H,); 7.40-7.30 (*m*, 7H); 7.15 (*dd*, J_1 = 8.3 Hz, J_2 = 2.2, 1H); 5.51 (*d*, *J* = 8.4 Hz, 1H); 4.44 (*dd*, J_1 = 10.2 Hz, J_2 = 3.8 Hz, 1H); 3.11 (*dd*, J_1 = 19.3 Hz, J_2 = 10.2 Hz, 1H); 2.80 (*dd*, J_1 = 19.3 Hz, J_2 = 3.8 Hz, 1H); 1.58 (*s*, 9H); 1.22 (*s*, 9H). ^{13}C NMR (100 MHz, $CDCl_3$): δ (ppm): 175.6, 175.4, 174.2, 150.1, 148.9, 141.6, 136.6, 131.8, 131.5, 130.5, 130.1, 129.2, 128.8, 128.6, 127.6, 126.3, 124.4, 124.1, 123.8, 116.6, 84.7, 57.0, 47.9, 35.7, 32.8, 31.3, 28.0. $[\alpha]_D^{20}$ +127.0 (*c* 2.00, $CHCl_3$). HRMS-ESI-ORBITRAP (+): calculated for $C_{33}H_{33}BrN_2NaO_5$ 639.1465, found 639.1454 $(M+Na)^+$. Calculated for $C_{33}H_{33}BrN_2KO_5$ 655.1204, found 655.1167 $(M+K)^+$.

(*M*)-(*R*)-*tert*-butyl-3-((*S*)-1-(4-bromo-2,5-di-tert-butylphenyl)-2,5-dioxopyrrolidin-3-yl)-2-oxo-3-phenylindoline-1-carboxylate (product **210**, Table 3.7—entry 13)

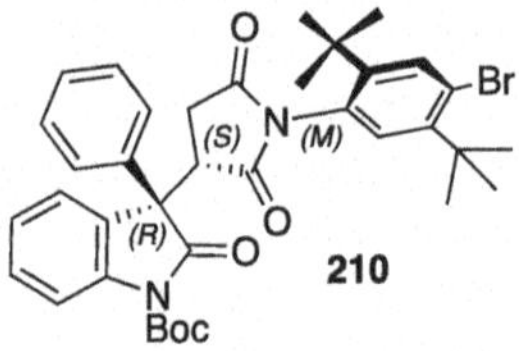

210

The title compound was obtained following the general procedure to furnish the crude product as a single of diastereoisomer (dr > 19:1). The crude mixture was purified by flash column chromatography (dichloromethane:hexane = 90:10) to give **210** in 50% yield, 0.100 mmol and 96% ee. HPLC analysis on a Daicel Chiralpak IC column: hexane/*i*-PrOH 80/20, flow rate 1 mL/min, λ = 254 nm: **210** τ_{major} = 5.7 min; τ_{minor} = 6.9 min. ^{1}H NMR (400 MHz, $CDCl_3$): δ (ppm): 8.06 (*d*, *J* = 8.3 Hz, 1H); 7.63 (*s*, 1H); 7.46 (*m*, 1H,); 7.41–7.25 (*m*, 7H); 5.95 (*s*, 1H); 4.45 (*dd*, J_1 = 10.4 Hz, J_2 = 4.3 Hz, 1H); 3.10 (*dd*, J_1 = 19.6 Hz, J_2 = 10.4 Hz, 1H); 2.82 (*dd*, J_1 = 19.6 Hz, J_2 = 4.3 Hz, 1H); 1.59 (*s*, 9H); 1.30 (*s*, 9H); 1.21 (*s*, 9H). ^{13}C-NMR (100 MHz, $CDCl_3$): δ (ppm): 175.7, 175.4, 174.4, 149.0, 146.9, 146.6, 141.5, 136.9, 136.0, 130.2, 129.2, 128.6, 128.3, 127.6, 126.4, 124.6, 124.3, 123.4, 116.5, 84.7,

56.6, 47.6, 36.1, 35.1, 32.8, 31.4, 29.6, 28.1. $[\alpha]_D^{20}$ +116.9 (*c* 1.00, $CHCl_3$). HRMS-ESI-ORBITRAP (+): calculated for $C_{37}H_{41}BrN_2NaO_5$ 695.2091, found 695.2083 $(M+Na)^+$. Calculated for $C_{37}H_{41}BrN_2KO_5$ 711.183, found 711.1938 $(M+K)^+$.

(*M*)-(*R*)-*tert*-butyl-3-((*S*)-1-(2,5-di-*tert*-butylphenyl)-2,5-dioxopyrrolidin-3-yl)-2-oxo-3-phenylindoline-1-carboxylate (product **211**, Table 3.7—entry 14)

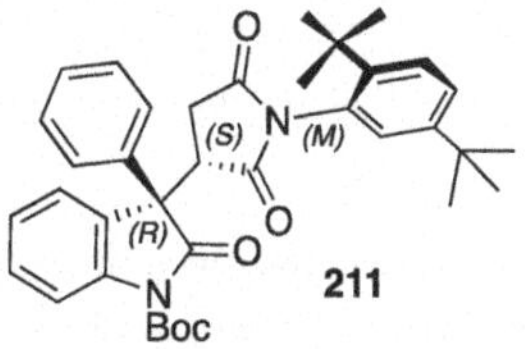

211

The title compound was obtained following the general procedure to furnish the crude product as a single of diastereoisomer (dr > 19:1). The crude mixture was purified by flash column chromatography (dichloromethane:hexane = 80:20) to give **211** in 43% yield, 0.086 mmol and > 99% ee. HPLC analysis on a Daicel Chiralpak AD-H column: hexane/*i*-PrOH 80/20, flow rate 0.8 mL/min, $\lambda = 254$ nm: **211** $\tau_{major} = 6.1$ min; $\tau_{minor} = 5.2$ min. ^{1}H NMR (400 MHz, $CDCl_3$): δ (ppm): 8.06 (*d*, $J = 8.3$ Hz, 1H); 7.50 (*m*, 1H,); 7.44–7.30 (*m*, 8H); 7.26 (*m*, 1H,); 5.81 (*d*, $J = 2.1$ Hz, 1H); 4.44 (*dd*, $J_1 = 10.2$ Hz, $J_2 = 3.8$ Hz, 1H); 3.10 (*dd*, $J_1 = 19.4$ Hz, $J_2 = 10.2$ Hz, 1H); 2.81 (*dd*, $J_1 = 19.4$ Hz, $J_2 = 3.8$ Hz, 1H); 1.58 (*s*, 9H); 1.22 (*s*, 9H); 1.14 (*s*, 9H). ^{13}C-NMR (100 MHz, $CDCl_3$): δ (ppm): 175.9, 175.8, 174.4, 150.1, 149.0, 144.4, 141.6, 136.9, 130.2, 129.1, 129.0, 128.5, 128.1, 127.6, 127.0, 126.8, 126.4, 124.4, 123.8, 116.4, 84.6, 56.8, 47.7, 35.1, 34.1, 32.8, 31.6, 31.1, 28.0. $[\alpha]_D^{20}$ +128.3 (*c* 1.00, $CHCl_3$). HRMS-ESI-ORBITRAP (+): calculated for $C_{37}H_{42}N_2NaO_5$ 617.2986, found 617.2979 $(M+Na)^+$. Calculated for $C_{37}H_{42}N_2KO_5$ 633.2725, found 633.2727 $(M+K)^+$.

(*M*)-(*R*)-*tert*-butyl-3-((*S*)-1-(2-(*tert*-butyl)-5-nitrophenyl)-2,5-dioxopyrrolidin-3-yl)-2-oxo-3-phenylindoline-1-carboxylate (product **212**, Table 3.7—entry 15)

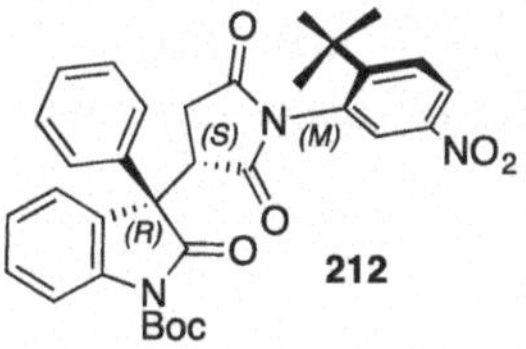

212

The title compound was obtained following the general procedure to furnish the crude product as a single of diastereoisomer (dr 19:1). The crude mixture was purified by flash column chromatography (hexane:ethyl acetate = 75:25) to give **212** in 81% yield, 0.162 mmol and 98% ee. HPLC analysis on a Daicel Chiralpak AD-H column: hexane/*i*-PrOH 80/20, flow rate 1 mL/min, $\lambda = 254$ nm: **212** $\tau_{major} = 17.6$ min; τ_{minor}

= 9.7 min. ^{1}H NMR (400 MHz, $CDCl_3$): δ (ppm): 8.08 (*m*, 2H,); 7.67 (*m*, 1H,); 7.63 (*d*, J = 9.0 Hz, 1H); 7.48 (*ddd*, $J_1 = J_2$ = 7.5 Hz, J_3 = 1.1 Hz, 1H); 7.38 (*m*, 6H); 6.56 (*d*, J = 2.5 Hz, 1H); 4.47 (*dd*, J_1 = 10.0 Hz, J_2 = 3.4 Hz, 1H); 3.16 (*dd*, J_1 = 19.4 Hz, J_2 = 10.0 Hz, 1H); 2.86 (*dd*, J_1 = 19.4 Hz, J_2 = 3.4 Hz, 1H); 1.58 (*s*, 9H); 1.27 (*s*, 9H). ^{13}C-NMR (100 MHz, $CDCl_3$): δ (ppm): 175.6, 175.3, 174.3, 155.9, 148.8, 146.5, 141.4, 136.5, 131.1, 130.9, 129.8, 129.3, 128.7, 127.7, 125.8, 125.6, 124.9, 124.3, 123.6, 116.8, 84.8, 57.2, 48.1, 36.4, 32.9, 31.3, 28.0. $[\alpha]_D^{20}$ +111.2 (*c* 1.00, $CHCl_3$). HRMS-ESI-ORBITRAP (+): calculated for $C_{33}H_{33}N_3NaO_7$ 606.2211, found 606.2202 $(M+Na)^+$. Calculated for $C_{33}H_{33}N_3KO_7$ 622.195, found 622.1919 $(M+K)^+$.

(*M*)-(*R*)-*tert*-butyl3-((*S*)-1-(5-(((benzyloxy)carbonyl)amino)-2-(tert-butyl)phenyl)-2,5-dioxopyrrolidin-3-yl)-2-oxo-3-phenylindoline-1-carboxylate (product **213**, Table 3.7—entry 16)

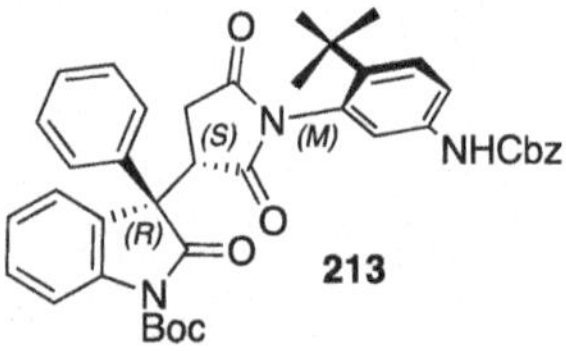

213

The title compound was obtained following the general procedure to furnish the crude product as a 6:1 mixture of diastereoisomers. The crude mixture was purified by flash column chromatography (hexane:ethyl acetate = 70:30) to give **213** in 94% yield, 0.188 mmol and 86% ee. HPLC analysis on a Daicel Chiralpak AD-H column: hexane/*i*-PrOH 80/20, flow rate 0.8 mL/min, λ = 254 nm: **213** τ_{major} = 35.8 min; τ_{minor} = 17.8 min. ^{1}H NMR (400 MHz, $CDCl_3$): δ (ppm): 8.04 (*d*, J = 8.7 Hz, 1H); 7.50–7.25 (*m*, 15H); 6.47 (*bs*, 1H); 5.73 (*bs*, 1H); 5.17 (*m*, 2H); 4.42 (*dd*, J_1 = 10.1 Hz, J_2 = 3.9 Hz, 1H); 3.07 (*dd*, J_1 = 19.4 Hz, J_2 = 10.1 Hz, 1H); 2.77 (*dd*, J_1 = 19.4 Hz, J_2 = 3.9 Hz, 1H); 1.56 (*s*, 9H); 1.20 (*s*, 9H). ^{13}C NMR (100 MHz, $CDCl_3$): δ (ppm): 175.7, 175.4, 174.4, 148.9, 141.3, 136.7, 136.7, 136.0, 134.8, 129.8, 129.1, 129.0, 128.6, 128.5, 128.4, 128.3, 128.1, 127.6, 127.2, 126.4, 125.0, 123.6, 119.9 (*bs*), 116.5, 84.6, 66.9, 56.8, 47.8, 35.1, 32.7, 31.5, 27.9. $[\alpha]_D^{20}$ +378.8 (*c* 2.00, $CHCl_3$). HRMS-ESI-ORBITRAP (+): calculated for $C_{41}H_{41}N_3NaO_7$ 710.2837, found 710.2832 $(M+Na)^+$. Calculated for $C_{41}H_{41}N_3KO_7$ 726.2576, found 726.2575 $(M+K)^+$.

References

1. Di Iorio N, Righi P, Mazzanti A, Mancinelli M, Ciogli A, Bencivenni G (2014) J Am Chem Soc 10:250
2. Di Iorio N, Champavert F, Erice A, Righi P, Mazzanti A, Bencivenni G (2016) Tetrahedron (invited paper) 5191 (special issue "Methods for controlling axial chirality")

3. Di Iorio N, Soprani L, Crotti S, Marotta E, Mazzanti A, Righi P, Bencivenni G (2017) Synthesis (invited paper) 1519
4. IUPAC (1997) Compendium of chemical terminology, 2nd ed. (the "Gold Book"). Oxford
5. Zeng XP, Cao ZY, Wang YH, Zhou F, Zhou J (2016) Chem Rev 7330
6. Borissov A, Davies TQ, Ellis SR, Fleming TA, Richardson MSW, Dixon DJ (2016) Chem Soc Rev 5474
7. Yamagata ADG, Datta S, Jackson KE, Stegbauer L, Paton RS, Dixon DJ (2015) Angew Chem Int Ed 4899
8. Curran DP, Qi H, Geib SJ, DeMello NC (1994) J Am Chem Soc 3131
9. Bencivenni G, Galzerano P, Mazzanti A, Bartoli G, Melchiorre P (2010) Proc Natl Acad Sci 20:642
10. Still WC, Kahn M, Mitra AJ (1978) J Org Chem 43:2923
11. Cassani C, Rapun RM, Arceo E, Bravo F, Melchiorre P (2013) Nat Protoc 325
12. Krafft ME, Wright JA (2006) Chem Commun 2977–2979
13. Wamg X, Reisinger CM (2008) B List J Am Chem Soc 130:6070–6071
14. Nolwenn JAM (2006) B List J Am Chem Soc 128:13368–13369
15. Bergman R, Magnusson G (1986) J Org Chem 51:212–217
16. Hadjiarapoglou L, Klein I, Spitzner D, de Meijere A (1996) *Synthesis* 525–528
17. Neumann JJ, Rakshit S, Dröge T, Glorius F (2009) Angew Chem Int Ed 48:6892–6895
18. Miller C, Hoyle C, Jönsson E (1998) PCT WO 98/54:134
19. Hamashima Y, Suzuki T, Takano H, Shimura Y, Sodeoka M (2005) J Am Chem Soc 10:164
20. Zhu XL, Xu JH, Cheng DJ, Zhao LJ, Liu XY, Tan B (2014) Org Lett 2192

Chapter 4
Direct Catalytic Synthesis of $C(Sp^2)$–$C(Sp^3)$ Atropisomers with Simultaneous Control of Central and Axial Chirality

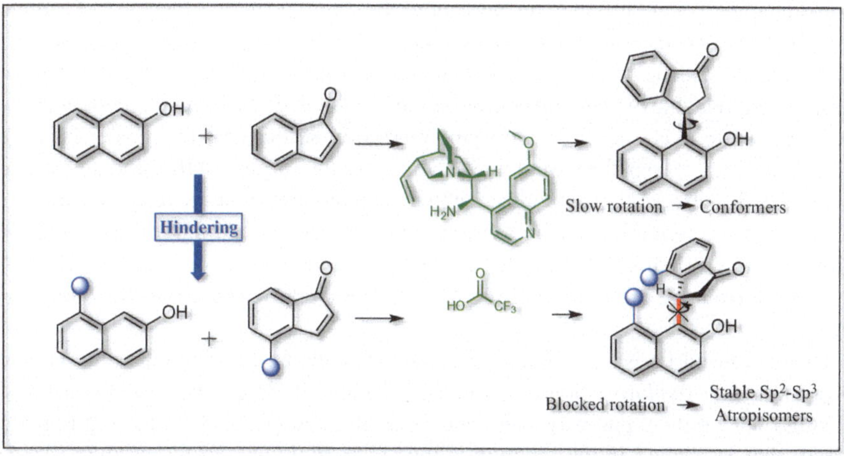

Parts of this chapter were adapted from Di Iorio et al. [1] with permission from ORGANIC LETTERS, copyright 2017 American Chemical Society.

Original content can be found at http://pubs.acs.org/doi/10.1021/acs.orglett.7b03415.

4.1 Forging a Stereogenic Axis

In the previous chapter, we discussed about the synthesis of axially chiral compounds using an indirect method to generate a stereogenic axis. The key point was to break the

---

Di Iorio et al. [1].

N. Di Iorio, *New Organocatalytic Strategies for the Selective Synthesis of Centrally and Axially Chiral Molecules*, Springer Theses,
https://doi.org/10.1007/978-3-319-74914-3_4

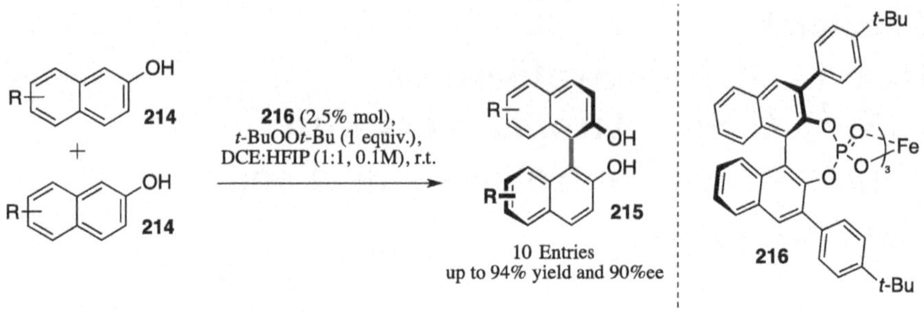

Reaction 4.1 Direct generation of a stereogenic axis in the ironphosphate-catalyzed homocoupling of naphthols

symmetry of a molecule and "reveal" the blocked rotation of a single sp^2–sp^2 bond.[1] In this chapter, we are going to discuss about the direct forming of a new C–C single bond that is itself the stereogenic axis. Naturally there is not a better approach for the synthesis of atropisomers and each one has its own advantages. Overall the most difficult feat to achieve is the formation of the hindered single bond that will become the stereogenic axis, therefore a desymmetrization strategy will be easier because the stereoselective transformation does not involve the forging of the hindered single bond. As a drawback, this method can only be applied to symmetric reagents whereas a catalytic stereoselective reaction virtually has no substrate limitations but because it involves the formation of a very hindered single bond is more difficult to realize. A very recent example of this kind of strategy has been reported by Toste (Reaction 4.1) [6].

Using a chiral phosphate ligand, the authors made an iron complex that is able to promote the oxidative homocoupling of 2-naphothols to give enantioenriched BINOLs via a radical pathway. It is easy to see that the new C–C bond (highlighted in red) that is formed in the reaction is the stereogenic axis in the product. Aside from a few other and this [7–11], there are not many reports of stereoselective catalytic generation of chiral axes and all these few examples show the formation of an sp^2–sp^2 axis. Although $C(sp^2)$–$C(sp^3)$ atropisomerism is well known, examples of systems presenting this kind of chirality are only focused on theoretical aspects such as conformational behavior and rotational barriers investigated by dynamic NMR and HPLC [12–17]. This is probably because of the drastic conditions needed to prepare these substrates that often do not allow for their catalytic synthesis. To the best of our knowledge there are no examples of catalytic, stereoselective formation of a $C(sp^2)$–$C(sp^3)$ chiral axis, therefore the aim of this work is to realize such transformation for the first time.

[1]For other examples of indirect generation of stereogenic axes see: Refs. [2–5].

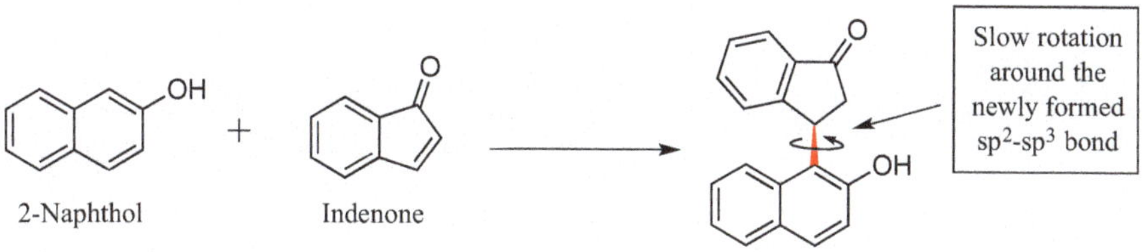

Fig. 4.1 Formation of the quasi-atropisomeric products

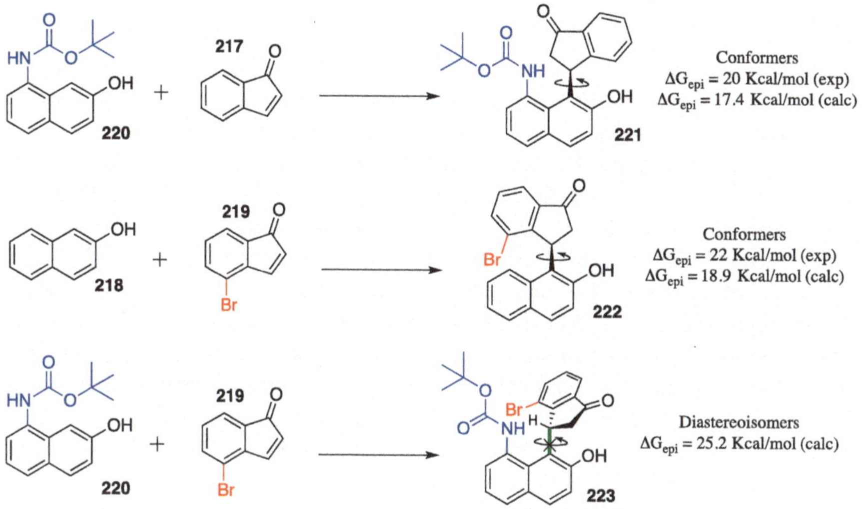

Scheme 4.1 Strategic hindrance on the reagents

4.2 Results and Discussion

We started thinking about this possibility after a paper by my research group was published [18] were they reported the FC reaction of naphthols and iminium ion-activated ketones. They noticed that performing the reaction with 2-naphthols and indenones gave the product as a mixture of conformers due to a slow rotation around the newly formed C(sp^2)–C(sp^3) bond (Fig. 4.1).

The rotational barrier of the highlighted bond is 17.8 kcal/mol which is not enough to generate atropisomeric compounds,[2] so we thought that hindering some strategic positions on the reagents would allow us to form a stable (sp^2)–(sp^3) stereogenic axis (Scheme 4.1).[3]

We started by hindering position 8 on the naphthol reagent and observed a greater barrier to rotation of 20 kcal/mol. When comparing this value with the calculated one

[2]For a stable atropisomer (at 25 °C) the rotational barrier should be higher than 25 kcal/mol

[3]Compound **221** shows fast degradation in DMSO and determination of the rotational barrier was impossible so the value is calculated for the corresponding OAc structure

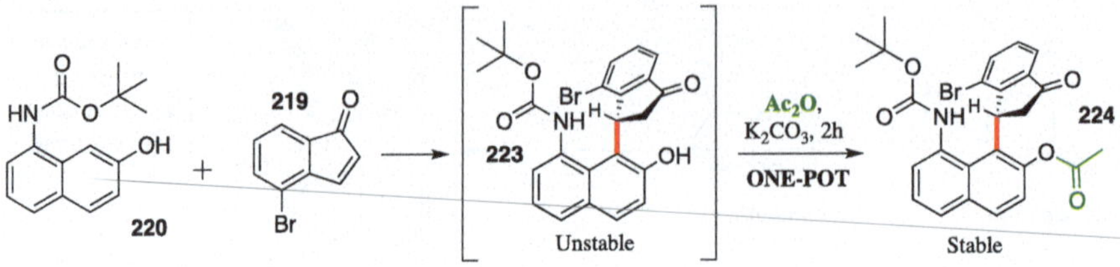

Reaction 4.2 One-pot acetylation of FC products

of 17.4 kcal/mol we see calculations underestimate it by approximately 3 kcal/mol which is nearly the same difference that we observed when we hindered position 4 on the indenone. This time the barrier (22 kcal/mol) is slightly higher but still not enough so we realized that to form a stable axis we needed to hinder both strategic positions. In this last case, we observed the formation of a single stable diastereoisomer **223** (*anti-periplanar*) whose calculated barrier was 25.2 kcal/mol. Unfortunately, we could not determine the barrier to rotation experimentally because the GS energy of the two diastereoisomers is very different (3.9 kcal/mol by calculations) meaning that the thermal equilibrium at a reasonable temperature will be completely shifted towards the more stable of the two diastereoisomers and to populate the higher energy GS very high temperatures are required that were beyond the physical possibilities of our instrumentations. Nevertheless, considering the calculated values of the two previous cases, it is reasonable to assume that the actual barrier is higher than 28 kcal/mol and that we formed a stable C(sp^2)–C(sp^3) axis.[4]

After setting the conditions to generate a stable axis, we had to deal with the general instability manifested by all the products obtained up to this point, so we figured that protecting the free OH group would improve the stability of the products and we found a one-pot procedure to protect the alcohol with an acetyl group without loss of optical purity (Reaction 4.2).

With this quick and reliable one-pot protection we could obtain stable products and move on to investigate the optimal reaction conditions (Table 4.1).[5]

We started with the solvent screening and obtained the best result in Chlorobenzene (Table 4.1—entry 5) so we verified that a different temperature, stoichiometry or concentration (Table 4.1—entries 9–12) did not affect the reaction in a positive way and moved on to the acid cocatalyst screening. We observed good results when strong aliphatic carboxylic acids were used with TFA (**229**) being the best one overall (Table 4.1—entry 19). Once we set the optimal conditions, we explored the scope of this atroposelective transformation reacting different indenones and naphthols together (Table 4.2).

[4]Also the experimental barriers reported in references [12–17] are very high (>30 kcal/mol) for compounds similar to ours.

[5]In Ref. [18], catalyst **140** had already been identified as the best amine so we did not perform an amine screening

Table 4.1 Screening of reaction conditions

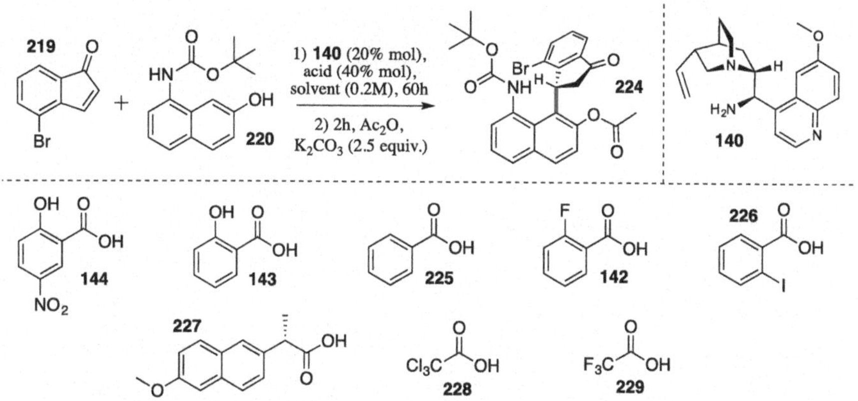

Entry	Solvent	Acid	Yield	d.r.	ee
1	PhBr	**144**	44%	>19:1	62%
2	Toluene	**144**	40%	>19:1	66%
3	PhF	**144**	57%	>19:1	64%
4	THF	**144**	10%	>19:1	78%
5	PhCl	**144**	53%	>19:1	80%
6	MTBE	**144**	<1%	>19:1	77%
7	DCM	**144**	60%	>19:1	60%
8	EtOAc	**144**	2%	>19:1	55%
9[a]	PhCl	**144**	63%	>19:1	77%
10[b]	PhCl	**144**	80%	>19:1	67%
11[c]	PhCl	**144**	63%	>19:1	70%
12[d]	PhCl	**144**	70%	>19:1	65%
13	PhCl	**143**	25%	>19:1	50%
14	PhCl	**225**	8%	>19:1	36%
15	PhCl	**142**	12%	>19:1	60%
16	PhCl	**226**	17%	>19:1	70%
17	PhCl	**227**	5%	>19:1	40%
18	PhCl	**228**	65%	>19:1	85%
19	PhCl	**229**	64%	>19:1	88%

[a]3 equiv. of naphthol; [b]40 °C; [c]0.1 M; [d]0.4 M

Table 4.2 Scope of the atroposelective FC reaction

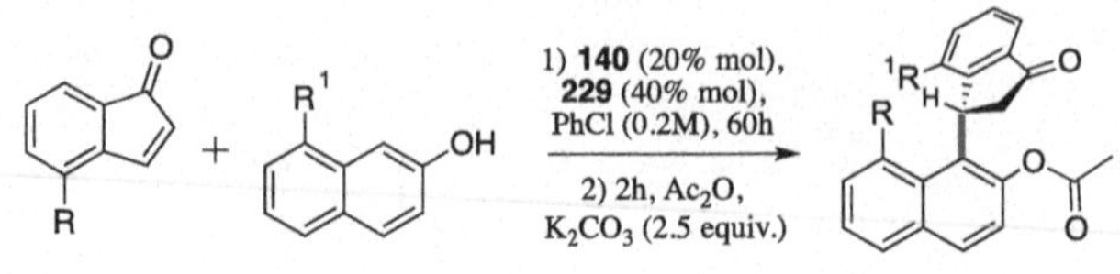

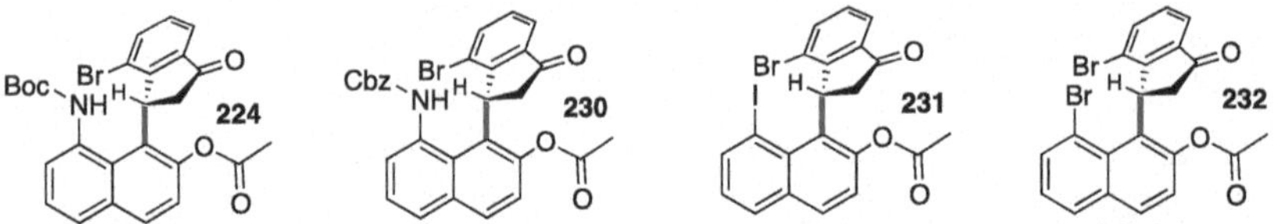

224: Yield 64%, d.r. >19:1, ee 88%

230: Yield 42%, d.r. >19:1, ee 82%

231: Yield 40%, d.r. >19:1, ee 97%

232: Yield 64%, d.r. >19:1, ee 97%

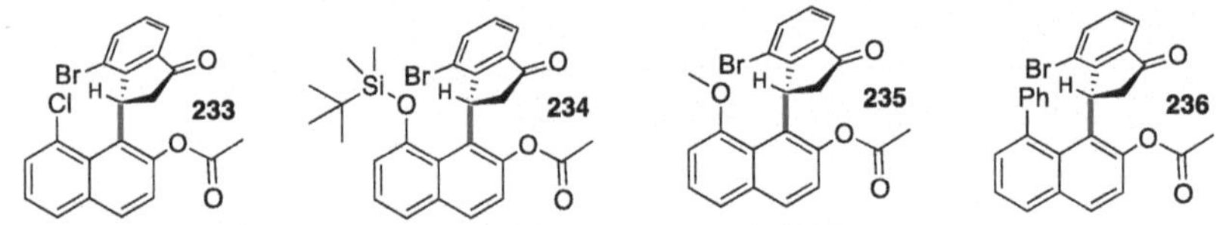

233: Yield 75%, d.r. >19:1, ee 96%

234: Yield 82%, d.r. >19:1, ee 94%

235: Yield 93%, d.r. >19:1, ee 90%

236: Yield 58%, d.r. >19:1, ee 94%

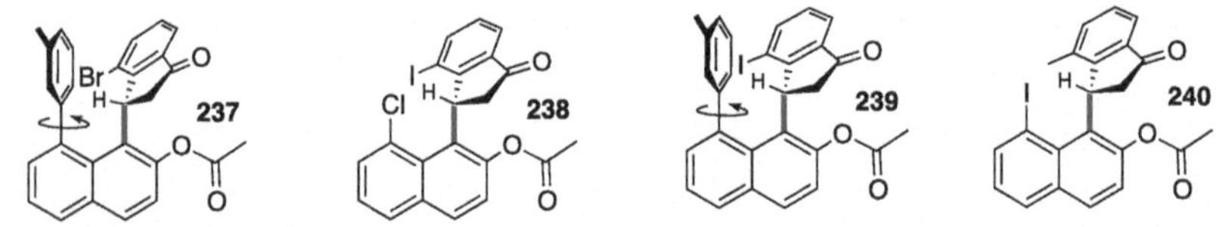

237: Yield 78%, d.r. >19:1, ee 95%

238: Yield 35%, d.r. >19:1, ee 93%

239: Yield 35%, d.r. >19:1, ee 94%

240: Yield 47%, d.r. >19:1, ee 93%

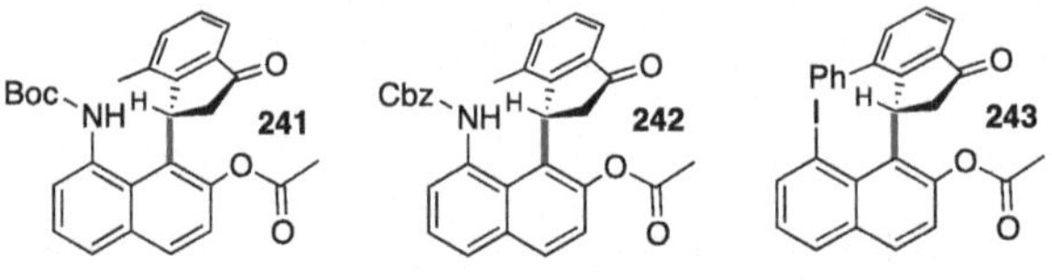

241: Yield 67%, d.r. >19:1, ee 86%

242: Yield 56%, d.r. >19:1, ee 83%

243: Yield 15%, d.r. >19:1, ee 98%

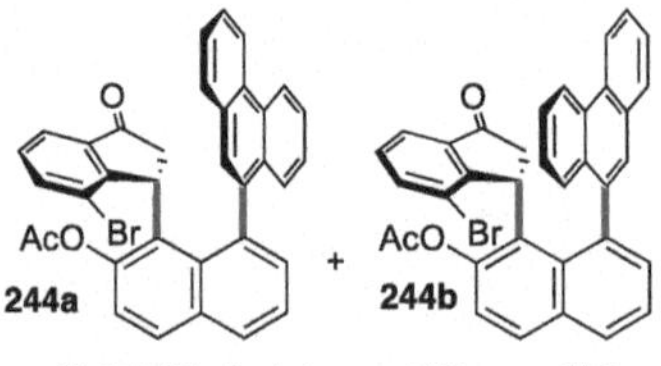

244a + **244b**

Yield 15%, d.r. 1:1, ee_{D1} 94%, ee_{D2} 65%

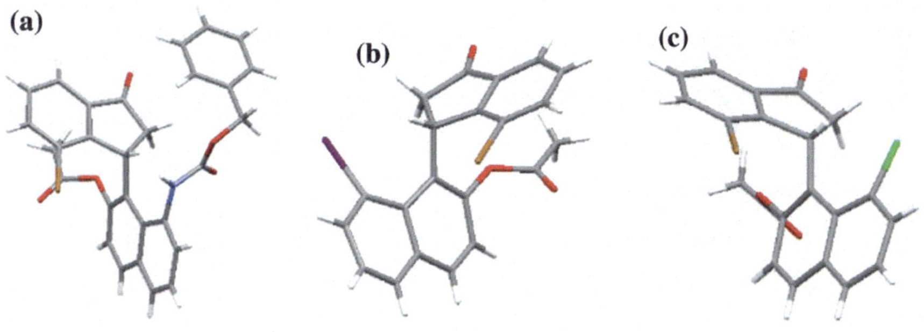

Fig. 4.2 XRD-derived structure of products. **a** 230; **b** ent-231; **c** 233

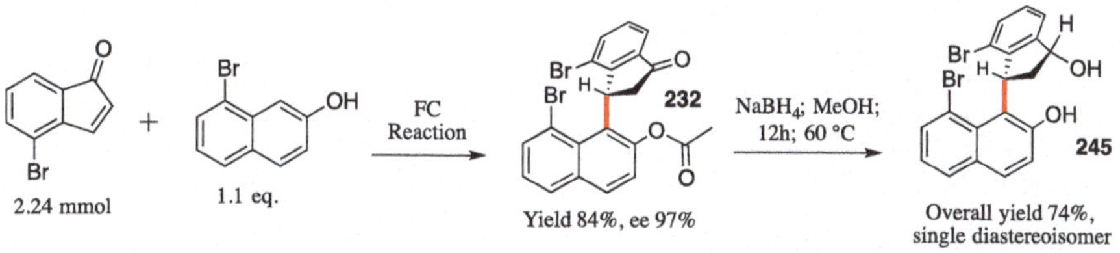

Reaction 4.3 Scale-up and application of the atroposelective FC reaction

All products were obtained as a single diastereoisomer in high enantioselectivity. Many naphthols were reacted with bromoindenone **219** with yields going from moderate to very good (Table 4.2—products **224–237**) and the same trend was observed for the reaction of different indenones (Table 4.2—products **238–243**). The absolute configuration was confirmed with single crystal XRD analysis on products **230**, *ent*-**231** and **233** and was found to be a *R,P* configuration (Fig. 4.2).

Interestingly products **237** and **239** were obtained as a 56:44 mixture of conformers due to a slow rotation around the (sp^2)–(sp^2) bond connecting the tolyl substituent with the naphthol structure, so we wondered if we could form a product bearing two stereogenic axis with this FC reaction. Pushed by curiosity we synthesized 8-phenantryl naphthol, reacted it with the indenone and obtained the product as a 1:1 mixture of diastereoisomers **244a**:**244b**. We then separated the two isomers via preparative HPLC and heated **244a** at 100 °C to determine the rotational barrier of the second axis that was found to be 29.9 kcal/mol.[6] Also by heating pure **244a** we obtained a mixture of **244a** and **244b** meaning that the diastereoisomerism is indeed due to the second axis which is not generated in the catalytic reaction.

Finally, we were easily able to scale up to ten times the reaction and we could perform a stereoselective derivatization with $NaBH_4$ of the product **232** in order to obtain the very interesting diol **245** in 73% yields over two steps (Reaction 4.3).

[6] See the experimental section for details.

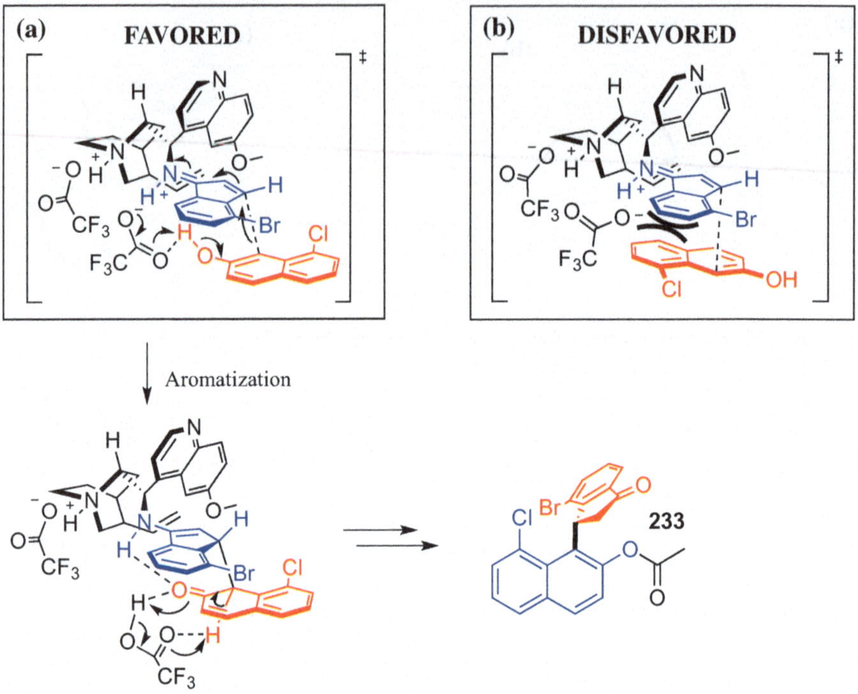

Fig. 4.3 Transition state for the FC reaction of hindered naphthols and indenones

4.3 Conclusions

Once again, we drew a reasonable transition state accounting for the selectivity observed where the catalytic system controls both the stereogenic center and axis (Fig. 4.3).

The *P* axial configuration can be obtained from two orientations of the naphthol and considering the impact that the acid cocatalyst has on the reaction (see Table 4.1), it is evident that the attack from the first orientation is favored (Fig. 4.3a). This is not only due to simple sterics, but also to the fact that the TFA anion is directly involved in a network of H-bonds and proton exchanges that accelerate the rate of the reaction and improve its selectivity.

In conclusion, we developed the first atroposelective synthesis of molecules possessing a C(sp^2)–C(sp^3) stereogenic axis that is the new C–C bond formed in the FC reaction. Strategic hindrance on the reagents together with primary amine catalysis were fundamental to make the axis stable and to achieve the high selectivity observed.

4.4 Experimental Section

4.4.1 General Information

All the NMR spectra were recorded on Inova 300 MHz, Gemini 400 MHz or Mercury 600 MHz Varian spectrometers for ^{1}H and 75, 100 and 150 MHz for ^{13}C respectively. The chemical shifts (δ) for ^{1}H and ^{13}C are given in ppm relative to the internal TMS standard or to the residual signals of $CHCl_3$. Coupling constants are given in Hz. Carbon multiplicities were determined by DEPT experiments. The following abbreviations are used to indicate the multiplicity: s, singlet; d, doublet; t, triplet; q, quartet; m, multiplet; bs, broad signal. Purification of reaction products was carried out by flash chromatography (FC) on silica gel (230–400 mesh) according to the method of Still [19]. Organic solutions were concentrated under reduced pressure on a Büchi rotary evaporator. High Resolution Mass spectra were obtained from the CIGS facilities of the University of Modena and Reggio Emilia on a G6520AA Accurate-Mass Q-TOF LC/MS instrument. X-ray data were acquired on a Bruker APEX-2 diffractometer. Chiral HPLC analysis was performed on an Agilent 1100-series instrumentation. Daicel Chiralpak AD-H, OD-H or AS-H columns with i-PrOH/hexane as the eluent were used. HPLC traces for the products were compared to *quasi* racemic samples prepared by mixing the two product antipodes obtained performing the reactions with catalyst **140** and the *pseudo*-enantiomer **30** separately. Optical rotations are reported as follows:$[\alpha]_{D}^{25}$(c in g per 100 mL, $CHCl_3$). All reactions were carried out in air and using undistilled solvents, without any precautions to exclude moisture unless otherwise noted.

4.4.2 General Procedure for the Synthesis of 4-Substituted Indenones

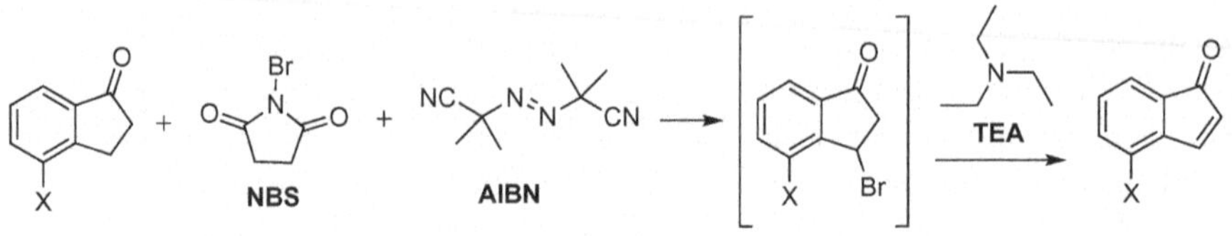

The appropriate indan-1-one (15 mmol, 3.2 g, 1 equiv.) was placed in a 100 mL round flask and suspended in CCl_4 (24 mL, 0.625 M) under magnetic stirring. The suspension was degased bubbling nitrogen with a needle for approximately 20 min at 35 °C (to better dissolve the indanone). After this time the bubbling was stopped and, paying attention to keep the flask always under nitrogen atmosphere, NBS (15 mmol, 2.7 g, 1 equiv.) and AIBN (1.5 mmol, 250 mg, 0.1 equiv.) were added and the solution was heated until boiling. After 3 h of reflux, a second portion of AIBN was added (1.5 mmol, 250 mg, 0.1 equiv.) and the solution was kept refluxing for 3 additional hours. At this point the heating was turned off and the solution was allowed to cool to room temperature, then the white precipitate (succinimide) was filtered and the residual liquid was concentrated to afford a brown/yellow solid that was suspended once more in Et_2O and treated with TEA (dropwise, 45 mmol, 6.6 mL, 3 equiv.) at 0 °C. The reaction was monitored by TLC and after completion, the TEA and its salts were removed by multiple washing with water, then the crude organic phase was made anhydrous with $MgSO_4$, concentrated and purified with flash column chromatography using an appropriate mixture of hexane and Et_2O. Both the crude and the pure indenone must be handled with extra care because they decompose easily in solution and in solid state. To avoid this issue the two synthetic steps were usually performed on the same day and after chromatography the pure product was stored at −18 °C, under argon atmosphere, in the dark.

1H-inden-1-one

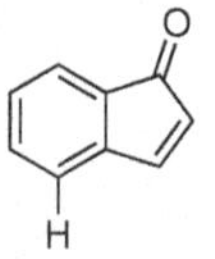

The reaction was carried out following the general procedure to furnish the crude product that was purified with flash column chromatography (hexane:Et_2O 95:5) to obtain the title compound as a yellow oil. ^{1}H NMR (300 MHz, $CDCl_3$): δ 7.55 (dd, 1H, $J_1 = 5.9$ Hz, $J_2 = 0.9$ Hz), 7.40 (m, 1H), 7.32 (m, 1H), 7.21 (ddd, 1H, $J_1 = J_2 = 7.9$ Hz, $J_3 = 0.9$ Hz), 7.04 (m, 1H), 5.89 (d, 1H, $J_1 = 5.9$ Hz). ^{13}C NMR (75 MHz, $CDCl_3$): δ 198.3, 149.7, 144.5, 133.6, 130.2, 129.0, 127.0, 122.5, 122.1.

4-bromo-1H-inden-1-one

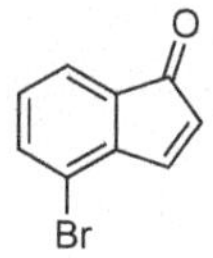

The reaction was carried out following the general procedure to furnish the crude product that was purified with flash column chromatography (hexane:Et_2O 95:5) to obtain the title compound as a bright yellow solid. ^1H NMR (300 MHz, $CDCl_3$): δ 7.67 (dd, 1H, J_1 = 6.0 Hz, J_2 = 0.9 Hz), 7.45 (dd, 1H, J_1 = 8.2 Hz, J_2 = 0.9 Hz), 7.36 (ddd, 1H, J_1 = 7.0 Hz, J_2 = J_3 = 0.9 Hz), 7.12 (dd, 1H, J_1 = 8.2 Hz, J_2 = 7.0 Hz), 5.98 (d, 1H, J_1 = 6.0.Hz). ^{13}C NMR (75 MHz, $CDCl_3$): δ 197.3, 148.8, 144.4, 136.9, 132.2, 130.8, 128.0, 121.4, 117.0.

4-iodo-1H-inden-1-one

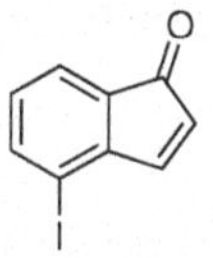

The reaction was carried out following the general procedure to furnish the crude product that was purified with flash column chromatography (hexane:Et_2O 95:5) to obtain 751 mg (40% yield) of the title compound as an orange solid. ^1H NMR (300 MHz, $CDCl_3$): δ 7.67 (dd, 1H, J_1 = 8.1 Hz, J_2 = 0.9 Hz), 7.52 (dd, 1H, J_1 = 6.0 Hz, J_2 = 0.9 Hz), 7.38 (ddd, 1H, J_1 = 7.0 Hz, J_2 = J_3 = 0.9 Hz), 6.99 (dd, 1H, J_1 = 8.1 Hz, J_2 = 7.0 Hz), 5.98 (d, 1H, J_1 = 6.0.Hz). ^{13}C NMR (75 MHz, $CDCl_3$): δ 197.9, 152.1, 148.8, 142.7, 132.2, 130.8, 128.0, 122.1, 90.7.

4-methyl-1H-inden-1-one

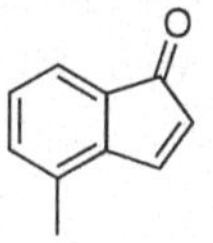

The reaction was carried out following the general procedure to furnish the crude product that was purified with flash column chromatography (hexane:Et_2O 95:5) to obtain the title compound as an orange viscous oil. ^1H NMR (300 MHz, $CDCl_3$): δ 7.67 (dd, 1H, J_1 = 6.0 Hz, J_2 = 0.9 Hz), 7.24 (dd, 1H, J_1 = 5.8 Hz, J_2 = 2.3 Hz), 7.12 (m, 2H), 5.84 (d, 1H, J_1 = 6.0.Hz), 2.28 (s, 3H). ^{13}C NMR (75 MHz, $CDCl_3$): δ 198.8, 148.0, 142.4, 135.4, 131.5, 130.2, 128.9, 126.3, 120.2, 16.8.

4-phenyl-1H-inden-1-one

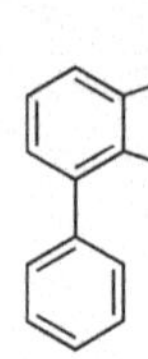

The reaction was carried out following the general procedure to furnish the crude product that was purified with flash column chromatography (hexane:Et_2O 90:10) to obtain the title compound as an orange-brown viscous oil. ^1H NMR (300 MHz, $CDCl_3$): δ 7.74 (dd, 1H, $J_1 = 6.0$ Hz, $J_2 = 0.9$ Hz), 7.53-7.38 (m, 7H), 7.31 (dd, 1H, $J_1 = 8.0$ Hz, $J_2 = 7.0$ Hz), 5.94 (d, 1H, $J_1 = 6.0$ Hz).

4.4.3 Synthesis of Naphthol Derivatives

Synthesis of *tert*-butyl (7-hydroxynaphthalen-1-yl)carbamate

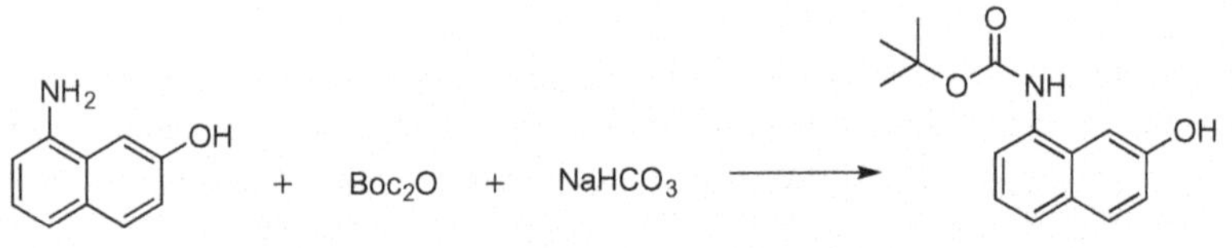

The commercially available 8-amino-2-naphthol (1.9 g, 12 mmol, 1 equiv.) and Boc_2O (2.64 g, 12.12 mmol, 1.01 equiv.) were dissolved in dry THF (30 mL, 0.4 M) under magnetic stirring and nitrogen atmosphere. After 4 days of refluxing the crude mixture was concentrated under vacuum and purified with flash column chromatography (hexane:EtOAc 2:1) to obtain 2.9 g of title compound as a pink-grey powder (93% isolated yield). ^1H NMR (300 MHz, $CDCl_3$): δ 7.77-7.47 (m, 3H), 7.25 (dd, 1H, $J_1 = J_2 = 7.4$ Hz), 7.12 (bs, 1H), 6.97 (dd, 1H, $J_1 = 8.7$ Hz, $J_2 = 1.4$ Hz), 6.64 (bs, 2H), 1.55 (s, 9H).

Synthesis of benzyl (7-hydroxynaphthalen-1-yl)carbamate

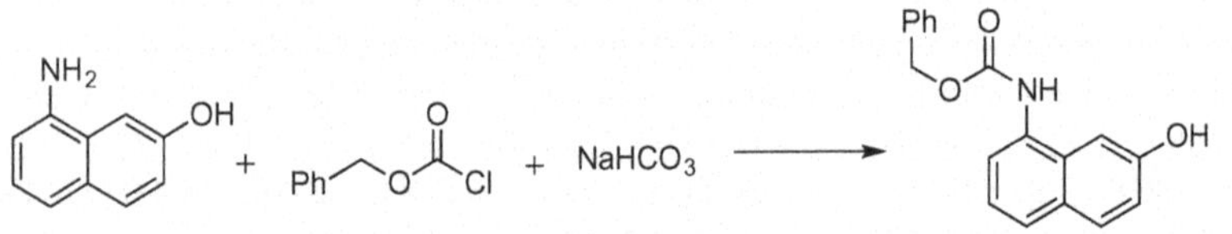

The commercially available 8-amino-2-naphthol (2.4 g, 15 mmol, 1 equiv.) and $NaHCO_3$ (1.26 g, 15 mmol, 1 equiv.) were dissolved in a 1:1 mixture of H_2O:THF (75 mL of mixture, 0.2 M) before adding benzylchloroformate (2.6 mL, 18 mmol, 1.2 equiv.) dropwise under magnetic stirring. The reaction was allowed to proceed overnight and was transferred in a separatory funnel then was diluted with 1 M HCl and extracted several times with Et_2O. The collected organic phases were reunited and made anhydrous over $MgSO_4$ before purification with flash column chromatography (hexane:EtOAc from 8:2 to 7:3) to obtain 3.28 g of title compound as a pink-grey powder (75% isolated yield). ^{1}H NMR (300 MHz, $CDCl_3$): δ 7.80-7.52 (m, 3H), 7.50-7.21 (m, 6 H), 7.06 (m, 2H), 6.74 (bs, 1H), 5.48 (bs, 1H), 5.24 (bs, 2H).

Synthesis of 8-iodonaphthalen-2-ol

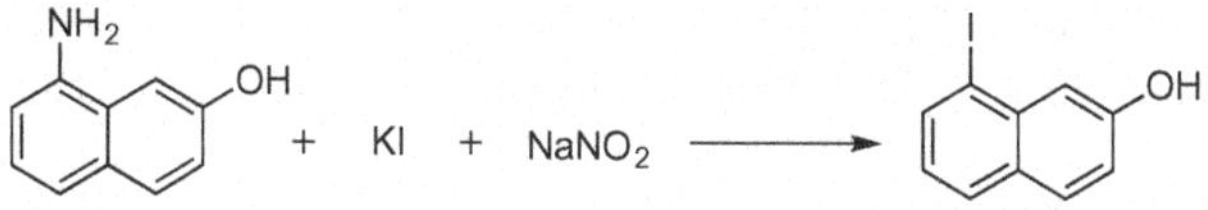

The commercially available 8-amino-2-naphthol (4 g, 25 mmol, 1 equiv.) was dissolved in a 1:2 mixture of THF:HCl 3 M (83 mL of mixture) and cooled to 0 °C under vigorous magnetic stirring. To this solution was added a second one of $NaNO_2$ (1.9 g, 28 mmol, 1.12 equiv.) in H_2O (8 mL) and after some stirring a third one made of KI (16.7 g, 100 mmol, 4 equiv.) in H_2O (12 mL). Both additions must be performed dropwise and the temperature must be monitored constantly so that it never rises higher than 4 °C. When the gas evolution stopped a final spoon of solid KI was added in the reaction vessel then the solution was diluted with EtOAc and washed several times with H_2O and brine. The organic layer was then made anhydrous over $MgSO_4$ concentrated and purified with flash column chromatography (hexane:EtOAc 8:2) to obtain 3.66 g of title compound as a white solid (55% isolated yield). ^{1}H NMR (300 MHz, $CDCl_3$): δ 8.02 (dd, 1H, J_1 = 7.4 Hz, J_2 = 1.2 Hz), 7.75 (dd, 1H, J_1 = 8.2 Hz, J_2 = 1.4 Hz), 7.69 (d, 1H, J_1 = 8.9 Hz), 7.44 (d, 1H, J_1 = 2.5 Hz), 7.13 (dd, 1H, J_1 = 8.9 Hz, J_2 = 2.5 Hz), 7.04 (dd, 1H, J_1 = 8.2 Hz, J_2 = 7.4 Hz), 5.17 (s, 1H).

Synthesis of 8-bromonaphthalen-2-ol and 8-chloronaphthalen

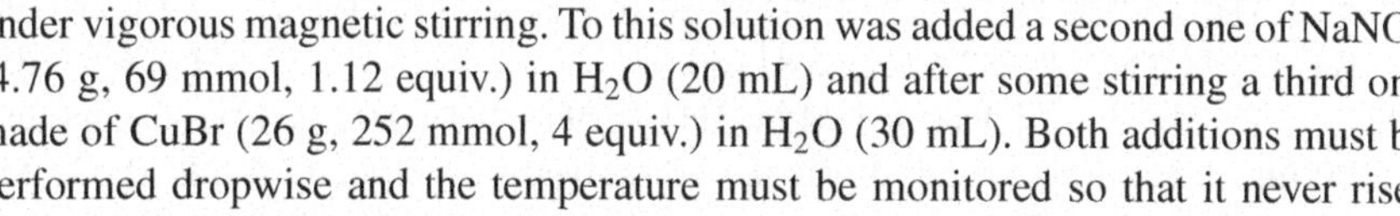

The commercially available 8-amino-2-naphthol (10 g, 62.5 mmol, 1 equiv.) was dissolved in a 1:2 mixture of THF:HCl 3M (150 mL of mixture) and cooled to 0 °C under vigorous magnetic stirring. To this solution was added a second one of $NaNO_2$ (4.76 g, 69 mmol, 1.12 equiv.) in H_2O (20 mL) and after some stirring a third one made of CuBr (26 g, 252 mmol, 4 equiv.) in H_2O (30 mL). Both additions must be performed dropwise and the temperature must be monitored so that it never rises higher than 4 °C. When the gas evolution stopped a final spoon of solid CuBr was added in the reaction vessel then the solution was diluted with EtOAc and washed several times with H_2O and brine. The organic layer was then made anhydrous over $MgSO_4$ concentrated and purified with flash column chromatography (hexane:EtOAc 8:2) to obtain a 85:15 mixture of **8-Br**:**8-Cl**. This mixture was most likely due to the HCl used for the diazotation and was unseparable with column chromatography regardless of the eluent we used so we employed preparative HPLC to isolate the two titled compounds. We used an chiralpak AD-H preparative column, hexane:IPA 95:5 as eluent and a 15 mL/min flow and were able to obtain the two title compound (**8-Cl** $R_f = 17.9$ min; **8-Br**: $R_f = 19.5$ min) separately. **8-Br**: ^{1}H NMR (300 MHz, $CDCl_3$): δ 7.73 (m, 3H), 7.55 (d, 1H, $J = 2.5$ Hz), 7.16 (m, 2H), 5.21 (s, 1H). ^{13}C NMR (75 MHz, $CDCl_3$): δ 154.7, 133.3, 130.6, 130.0, 127.7, 123.9, 120.9, 118.5, 109.2. **8-Cl**: ^{1}H NMR (300 MHz, $CDCl_3$): δ 7.75 (d, 1H, $J = 8.8$ Hz), 7.67 (d, 1H, $J = 8.2$ Hz), 7.53 (m, 2H), 7.27-7.10 (m, 2H), 5.37 (s, 1H). ^{13}C NMR (75 MHz, $CDCl_3$): δ 154.4, 132.1, 130.4, 130.1, 129.9, 126.9, 126.7, 123.4, 118.5, 106.5.

Synthesis of 8-((*tert*-butyldimethylsilyl)oxy)naphthalen-2-ol (2g) and 8-methoxynaphthalen-2-ol

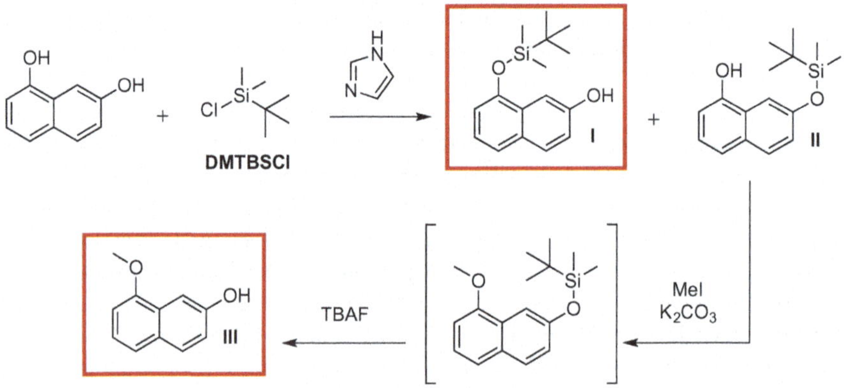

The commercially available 1,7-naphthalenediol (960 mg, 6 mmol, 1 equiv.), imidazole (857 mg, 12.6 mmol, 2.1 equiv) and DMTBSCl (882 mg, 5.88 mmol, 0.98 equiv.) were dissolved in DMF (12 mL, 0.5 M) and were left stirring for 2 h. After this time the reaction was complete so the mixture was diluted with Et_2O, the organic layer was washed several times with water, was made anhydrous over $MgSO_4$, concentrated and purified with flash column chromatography (hexane: Et_2O 80:20) to obtain a 1:2 mixture of **I**:**II**. These two isomers were isolated once again by preparative HPLC using a chiralpak AD-H column, hexane:IPA 90:10 as eluent and a 20 mL/min flow (**I** $R_f = 10.0$ min; **II**: $R_f = 4.7$ min). **I**: ^{1}H NMR (300 MHz, $CDCl_3$): δ 7.41 (d, 1H, $J = 8.9$ Hz), 7.27 (d, 1H, $J = 2.6$ Hz), 7.11 (d, 1H, $J = 8.3$ Hz), 6.90 (dd, 1H, $J_1 = 8.2$ Hz, $J_2 = 7.6$ Hz), 6.81 (dd, 1H, $J_1 = 8.9$ Hz, $J_2 = 2.6$ Hz), 6.57 (dd, 1H, $J_1 = 7.6$ Hz, $J_2 = 1.1$ Hz), 5.44 (s, 1H), 0.79 (s, 9H), 0.02 (s, 6H). Compound **II** (400 mg, 1.5 mmol, 1 equiv.) was then dissolved in acetone (2.4 mL, 0.625 M) together with K_2CO_3 (234 mg, 1.7 mmol, 1.13 equiv.) and MeI (145 μL, 2.3 mmol, 1.53 equiv.) and the mixture was refluxed for 3 h under magnetic stirring. After this time the heating was turned off and the solution was concentrated and flushed with Et_2O through a short plug of silica to remove K_2CO_3, then concentrated again and dissolved in THF (7.5 mL, 0.2 M) together with TBAF (905 mg, 4.5 mmol, 3 equiv.). After 1 h under magnetic stirring the reaction was complete so the crude was concentrated and purified with flash column chromatography (hexane:Et_2O 80:20) to afford compound **III**. **III**: ^1H NMR (400 MHz, $CDCl_3$): δ 7.70 (d, 1H, $J = 8.9$ Hz), 7.54 (d, 1H, $J = 2.5$ Hz), 7.36 (d, 1H, $J = 8.3$ Hz), 7.23 (dd, 1H, $J_1 = J_2 = 7.8$ Hz), 7.10 (dd, 1H, $J_1 = 8.9$ Hz, $J_2 = 2.6$ Hz), 6.78 (d, 1H, $J = 7.8$ Hz), 5.22 (s, 1H), 3.96 (s, 3H).

4.4.4 General Procedure for the Synthesys of 8-Arylnaphthalen-2-Ol

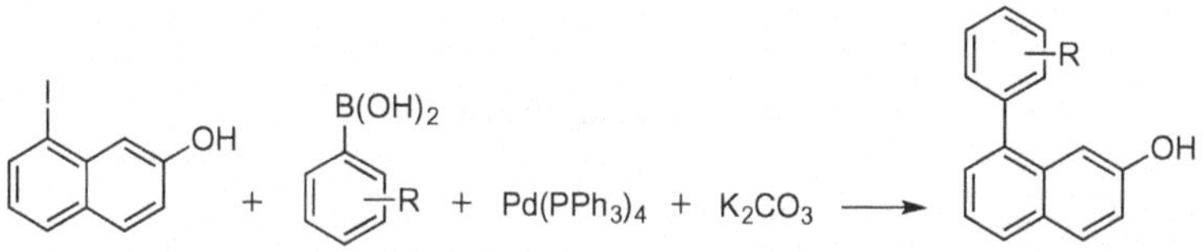

8-Iodonaphthol (1 equiv.) and boronic acid (1.1 equiv.) were placed in the reaction vessel and dissolved in a 1:1:2 solution of 2 M $K_2CO_{3(acq)}$:EtOH:toluene in order to obtain a concentration of naphthol ~0.175M. This mixture was then degassed by sonication under mild vacuum (water pump) for approximately 30 min before adding the tetrakis(triphenylphosphine)palladium and refluxing under magnetic stirring and nitrogen atmosphere. After 6 h the heating was turned off, the toluene was evaporated under vacuum, EtOAc was added and this mixture was washed several times with water. The organic layer was then made anhydrous on $MgSO_4$ and the crude product was purified with flash column chromatography.

8-Phenylnaphthalen-2-ol

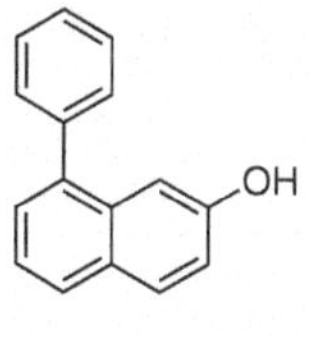

The reaction was carried out following the general procedure to furnish the crude product that was purified with flash column chromatography (hexane:Et_2O 80:20) to obtain 920 mg (90% yield) of title compound as a light-brown viscous oil. ^1H NMR (400 MHz, $CDCl_3$): δ 7.70 (m, 2H), 7.41-7.27 (m, 7H), 7.13 (d, 1H, $J_1 = 2.5$ Hz), 7.00 (dd, 1H, $J_1 = 8.9$ Hz, $J_2 = 2.5$ Hz), 5.39 (s, 1H). ^{13}C NMR (100 MHz, $CDCl_3$): δ 153.4, 140.8, 138.6, 132.7, 130.2, 129.8, 129.2, 128.2, 127.5, 127.4, 127.1, 123.1, 117.4, 107.9.

8-Tolylnaphthalen-2-ol

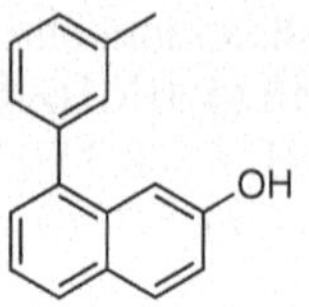

The reaction was carried out following the general procedure to furnish the crude product that was purified with flash column chromatography (hexane:Et_2O 80:20) to obtain 590 mg (75% yield) of title compound as a light-brown viscous oil. ^{1}H NMR (300 MHz, $CDCl_3$): δ 7.69 (m, 2H), 7.38-7.06 (m, 7H), 6.99 (dd, 1H, J_1 = 8.9 Hz, J_2 = 2.5 Hz), 5.45 (s, 1H), 2.31 (s, 3H). ^{13}C NMR (75 MHz, $CDCl_3$): δ 153.3, 140.7, 138.8, 137.9, 132.8, 130.5, 130.2, 129.2, 128.1, 127.8, 127.4, 127.3, 126.9, 123.1, 117.4, 108.0, 21.3.

8-Phenantrylnaphthalen-2-ol

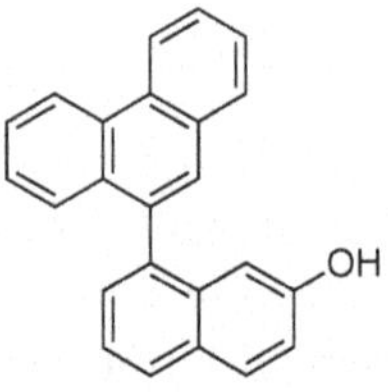

The reaction was carried out following the general procedure to furnish the crude product that was purified with flash column chromatography (hexane:Et_2O 80:20) to obtain 500 mg (85% yield) of title compound as a white solid. ^{1}H NMR (300 MHz, $CDCl_3$): δ 8.68 (m, 2H), 7.80 (m, 3H), 7.69 (s, 1H), 7.67-7.37 (m, 6H), 7.32 (ddd, 1H, J_1 = 8.1 Hz, J_2 = 6.8 Hz, J_3 = 1.2 Hz), 6.99 (dd, 1H, J_1 = 8.9 Hz, J_2 = 2.6 Hz), 6.62 (d, 1H, J_1 = 2.6 Hz), 4.70 (s, 1H). ^{13}C NMR (75 MHz, $CDCl_3$): δ 153.5, 137.1, 136.7, 134.0, 131.9, 131.6, 130.3, 130.2, 130.1, 128.9, 128.6, 128.4, 128.3, 127.8, 127.4, 126.9, 126.7, 126.6, 126.5, 123.2, 122.8, 122.6, 117.7, 108.3.

4.4.5 *Determination of the Barrier to Racemization Relative to the Naphthol-Phenantrene Stereogenic Axis for Compound 244*

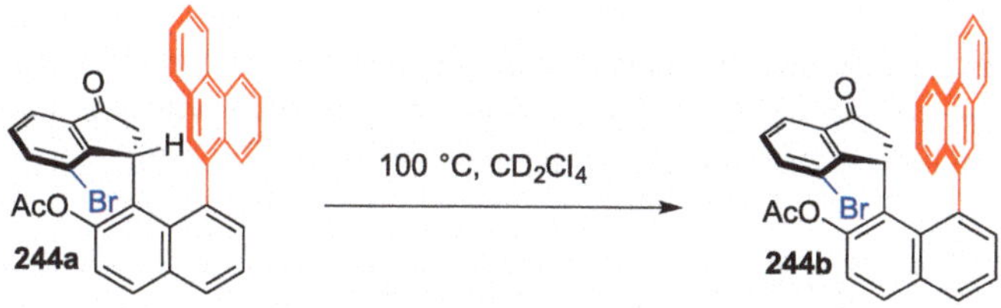

t (s)	x	ln(x - xeq)
0	1	-0.89648
900	0.98	-0.94674
1800	0.9606	-0.99804
3600	0.9363	-1.06624
7200	0.8757	-1.25983
10800	0.823	-1.46533
14400	0.7716	-1.71702
18000	0.7496	-1.84769
21600	0.7097	-2.13961
28800	0.6766	-2.46982
∞	0.592	

$\Delta G^{\neq} = 29.94$ kcal/mol

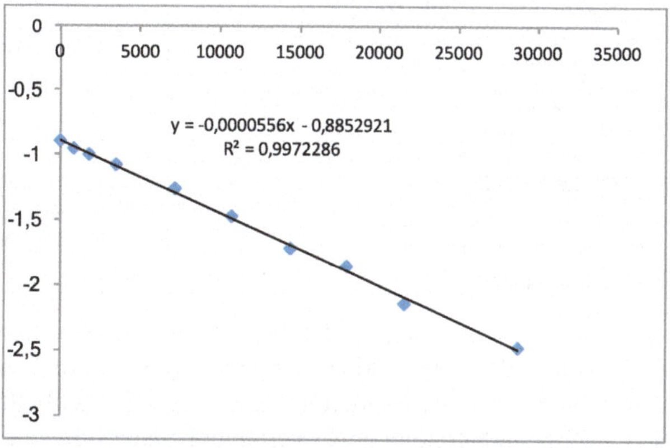

Kinetic measurement of the thermal equilibration of the naphthol-phenanthrene axis of the two diastereoisomers **244a–244b**. The sample was kept in CD_2Cl_4 at 100 °C while the NMR ratio measurements were obtained at 25 °C.

4.4.6 Experimental Determination of the Energy Barrier to Rotation and Estimated Value Through DFT Calculation of Compound 221

Ground states

In the case of **221**, the DFT calculations suggested an energy difference between the ap and sp conformation of 1.0 kcal/mol in agreement with the observation of both conformation in the ambient temperature NMR (DMSO-d6).

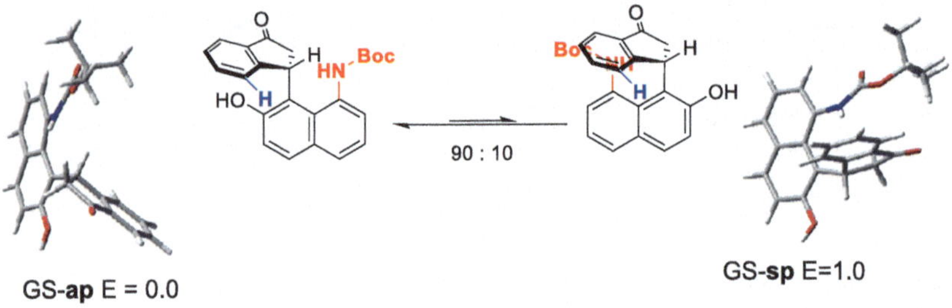

Transition states

For the interconversion of the most stable *ap*-conformer of **221** to the less stable *sp*-conformer through rotation along the C(sp^3)–C(sp^2) single bond, two transition states (TS) have been considered (Figure B). TS-1 is obtained when the –CH_2–C=O and C^4 of the indanone moiety overlap the NBoc substituent and the hydroxy group of naphthol respectively (violet arrow). The calculated barrier to rotation in this case is 17.4 kcal/mol. TS-2 is obtained when –CH_2–C=O and C^4 of the indanone moiety overlap the hydroxy group and the NBoc substituent of naphthol respectively (green arrow). The calculated barrier to rotation in this case is 18.8 kcal/mol. The optimized geometries shown were validated as real TS by frequency analysis, showing a single imaginary frequency corresponding to the rotation of the indanone ring.

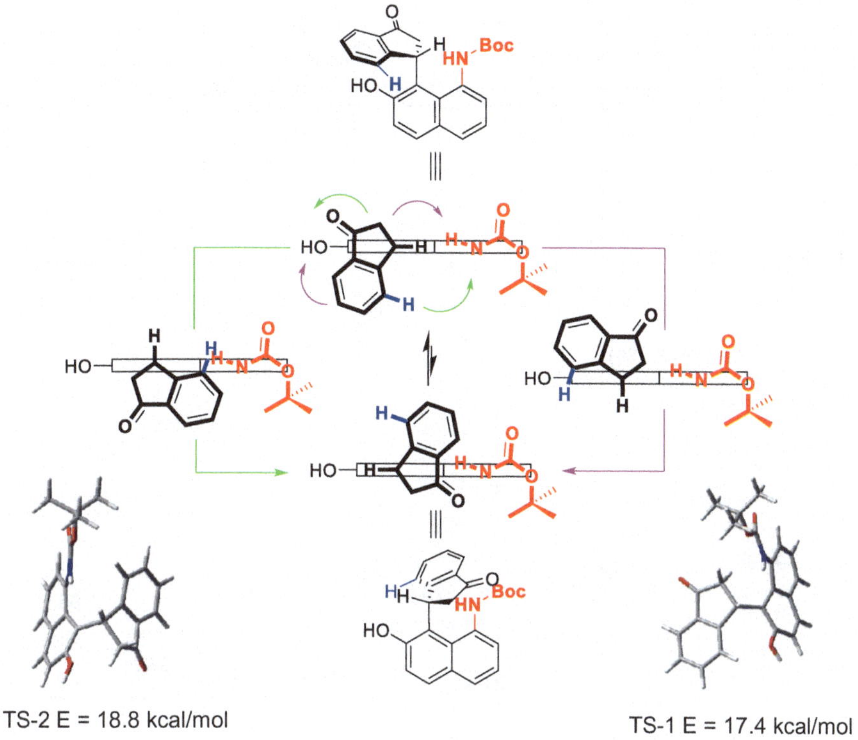

Experimental

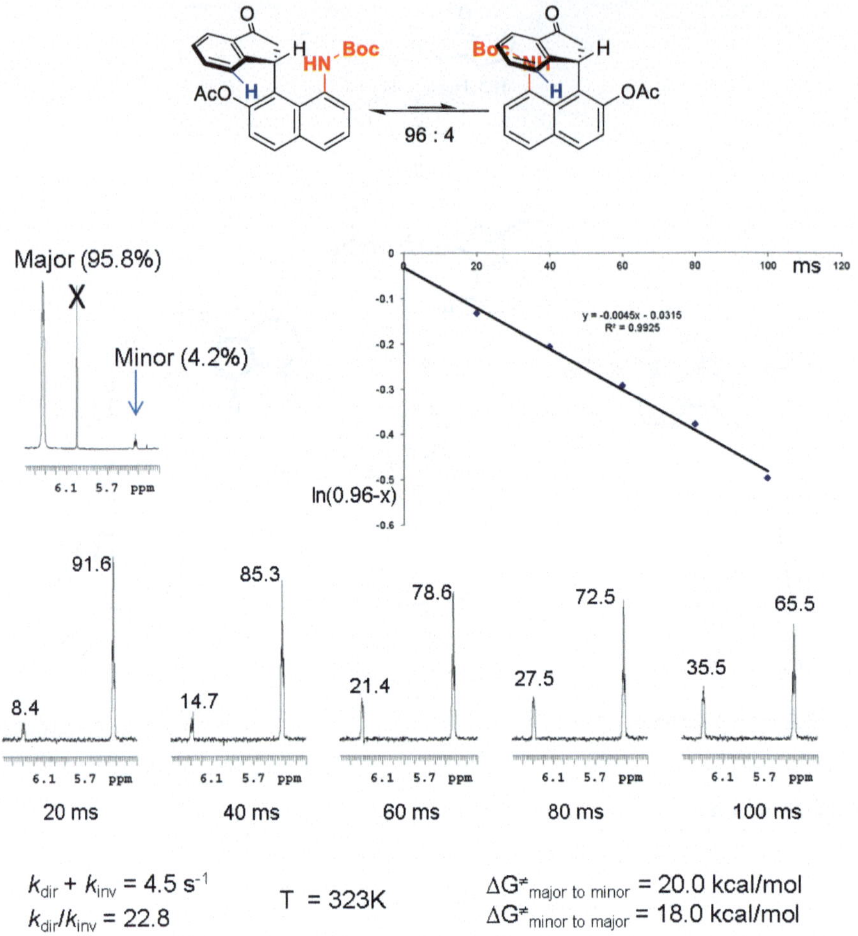

AcO
HN
Boc
OAc
96 : 4
Major (95.8%)
Minor (4.2%)
6.1 5.7 ppm
y = -0.0045x - 0.0315
R² = 0.9925
ms
ln(0.96-x)
91.6
8.4
85.3
14.7
78.6
21.4
72.5
27.5
65.5
35.5
20 ms
40 ms
60 ms
80 ms
100 ms
$k_{dir} + k_{inv} = 4.5\ s^{-1}$
$k_{dir}/k_{inv} = 22.8$
T = 323K
$\Delta G^{\neq}_{major\ to\ minor} = 20.0$ kcal/mol
$\Delta G^{\neq}_{minor\ to\ major} = 18.0$ kcal/mol

4.4.7 *Experimental Determination of the Energy Barrier to Rotation and Estimated Value Through DFT Calculation of Compound 222*

Ground states

In the case of the less hindered **222**, the DFT calculations suggested a very small energy difference between the *ap* and *sp* conformation, in agreement with the observation of both conformation in the ambient temperature NMR (DMSO-d6).

@@

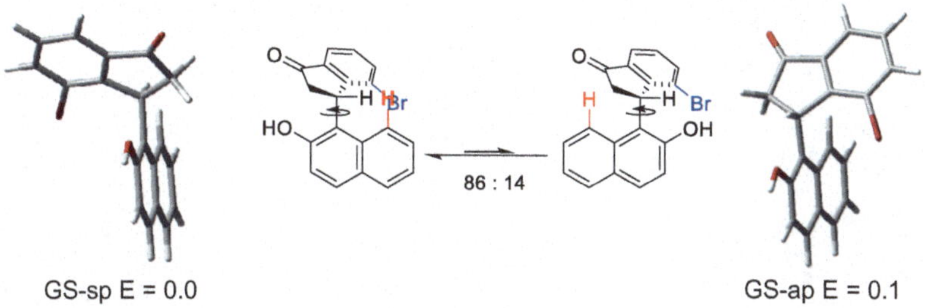

Transition states

For the interconversion of the most stable *ap*-conformer of **222** to the less stable *sp*-conformer through rotation along the C(sp^3)–C(sp^2) single bond, two transition states (TS) have been considered. TS-1 is obtained when the –CH_2–C=O and the bromine atom of the indanone moiety overlap the H^8 and the hydroxy group of naphthol respectively (violet arrow). The calculated barrier to rotation in this case is 22.0 kcal/mol. TS-2 is obtained when –CH_2–C=O and the bromine atom of the indanone moiety overlap the hydroxy group and the H^8 of naphthol respectively (green arrow). The calculated barrier to rotation in this case is 18.9 kcal/mol. The optimized geometries shown in Figure E were validated as real TS by frequency analysis, showing a single imaginary frequency corresponding to the rotation of the indanone ring.

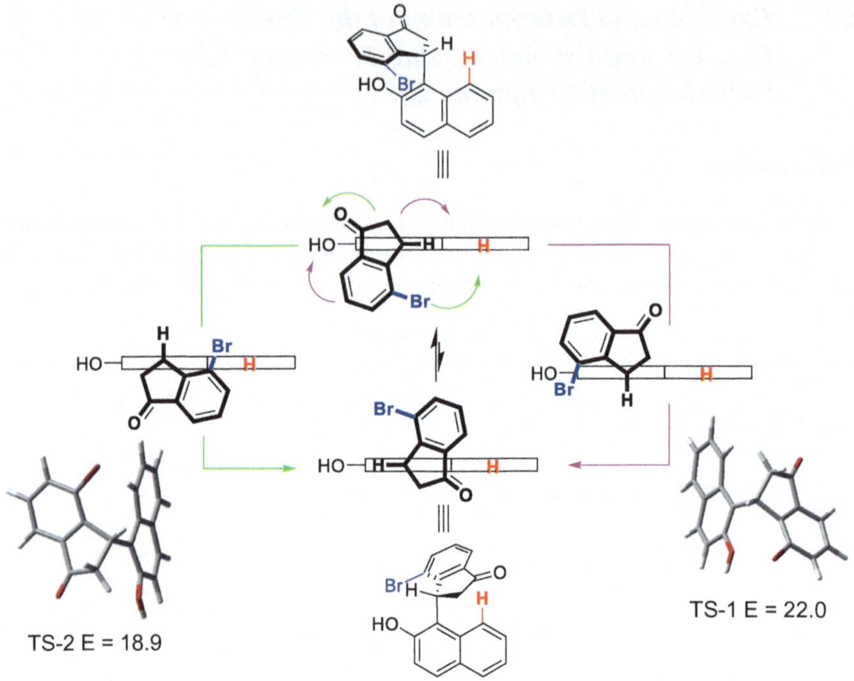

Experimental

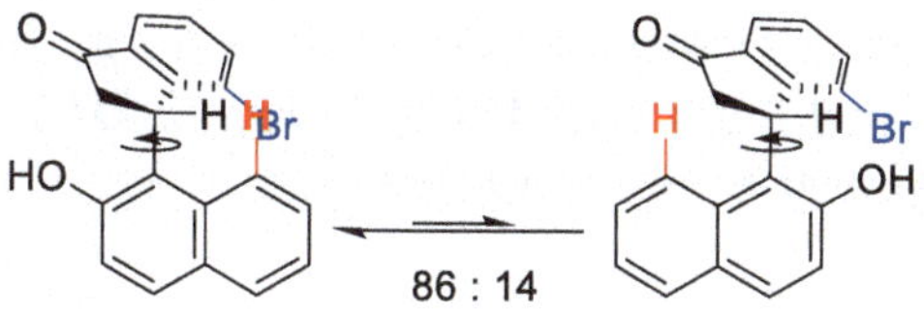

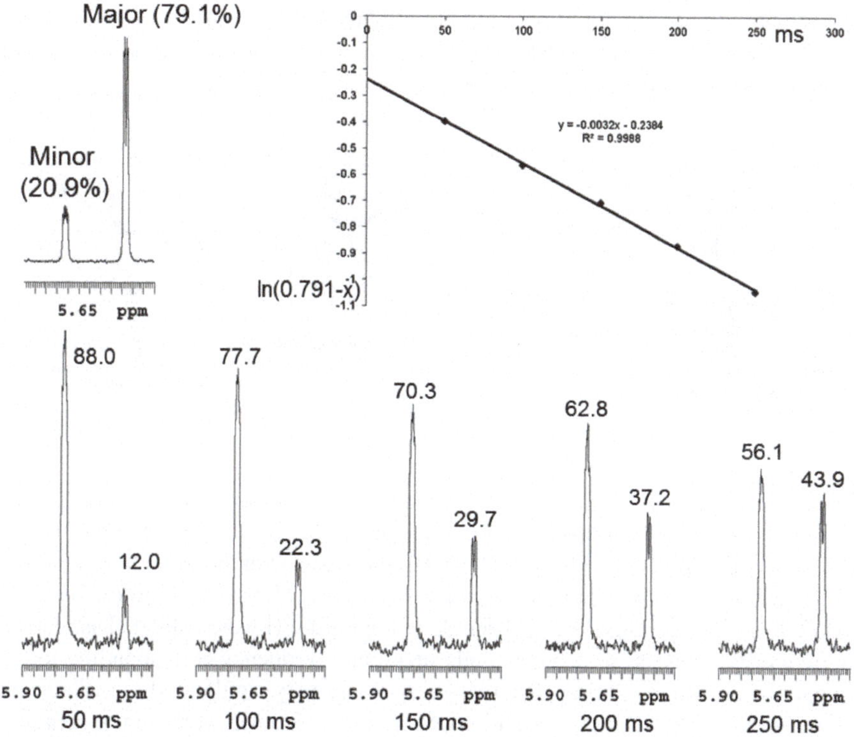

$k_{dir} + k_{inv} = 3.19\ s^{-1}$

$k_{dir}/k_{inv} = 3.78$

$T = 368K$

$\Delta G^{\neq}_{major\ to\ minor} = 22.0$ kcal/mol

$\Delta G^{\neq}_{minor\ to\ major} = 21.0$ kcal/mol

4.4.8 *Experimental Determination of the Energy Barrier to Rotation and Estimated Value Through DFT Calculation of Compound 224*

Ground states

The ground states of a model compound of **224** (where the OAc was replaced by OH to reduce computational times) were optimized using B3LYP and 6-31 g(d) basis set. The *ap*-conformation (CH of indanone close to NH) was found to be more stable than the sp (CH close to OH) by 4.0 kcal/mol in agreement with the observation of a single conformer in the ambient temperature NMR (DMSO-d6).

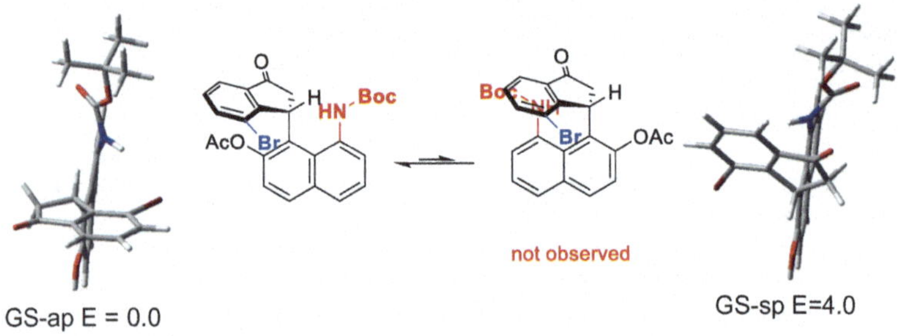

Transition states

Two diastereomeric transition states are conceivable for the conversion of the ap into the sp conformations. One correspond to the crossing of the bromine in position 4 of indenone over the NH (TS-1), while the second TS has the bromine over the OH (TS-2). The optimized geometries shown in Figure H were validated as real TS by frequency analysis, showing a single imaginary frequency corresponding to the rotation of the indanone ring. The two energies are very similar and the values are high enough to allow for the formation of stable atropisomers at room temperature. It should be noted that the substitution of the acetyl group with a hydrogen implies a slightly lower barrier to rotation of the indanone ring. For, the suggested barriers are lower with respect to the barriers of **224**.

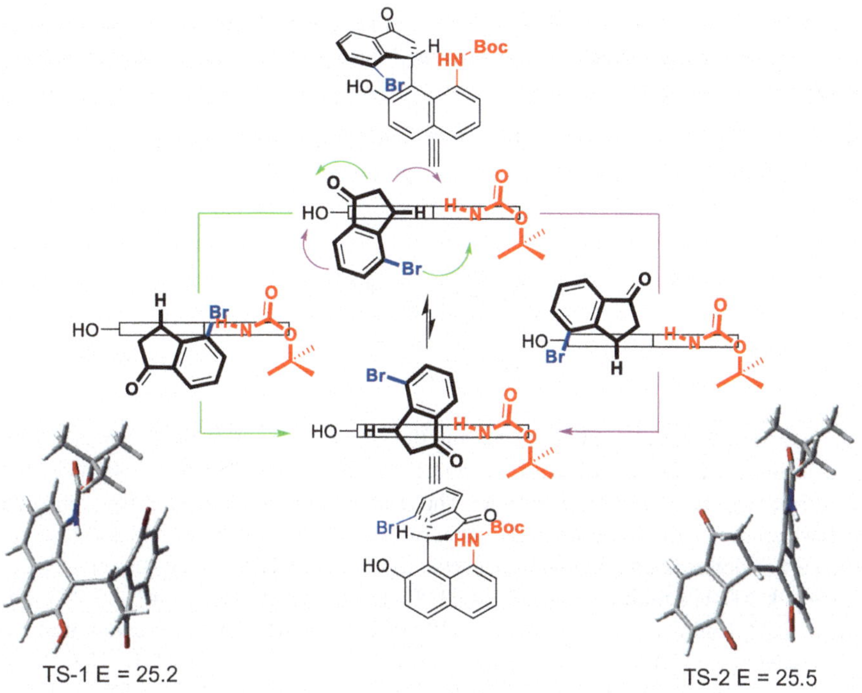

4.4.9 General Procedure for the Atroposelective FC Reaction

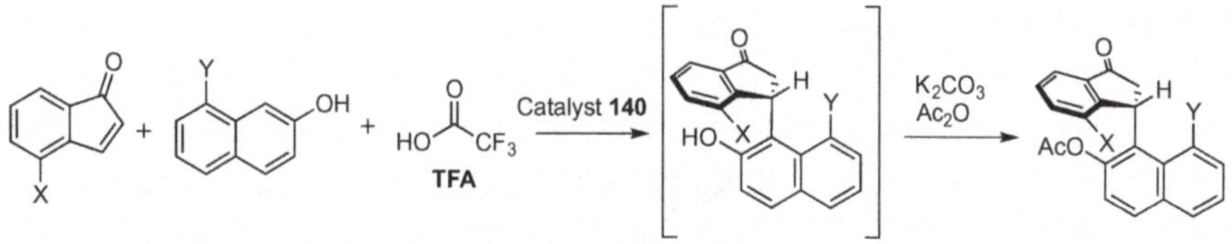

In an ordinary vial equipped with a Teflon-coated magnetic stir bar, catalyst **140** (13 mg, 0.04 mmol, 0.2 equiv.) was dissolved in 1 mL of a freshly prepared 9 mg/mL solution of TFA (9 mg, 0.08 mmol, 0.4 equiv.) in chlorobenzene (0.2 M). At this point the vial was covered with aluminium foil to shield from light and then the indenone (0.2 mmol, 1 equiv.) and the naphthol (0.22 mmol, 1.1 equiv.) were added respectively. After 56–60 of stirring, K_2CO_3 (69 mg, 0.5 mmol, 2.5 equiv.) and acetic anhydride (0.5 mL) were added and the solution was allowed to stir 2 additional hours before flushing it through a short silica plug with a 1:1 mixture of DCM:EtOAc to remove the catalyst. At this stage the crude product was first concentrated to perform a ^{1}H-NMR analysis to determine the d.r. and then purified with flash column chromatography. Finally the ee% was determined through HPLC on a chiral stationary phase.

(S)-8-((tert-butoxycarbonyl)amino)-1-(3-oxo-2,3-dihydro-1H-inden-1-yl)naphthalen-2-yl acetate (Scheme 4.1—product **221**)

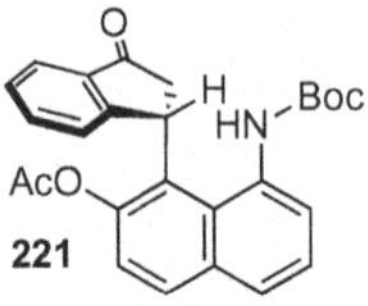

221

The reaction was carried out following the general procedure using catalyst **140**, acid cocatalyst **144** in dry toluene to furnish the crude product **221** as a 96:4 mixture of conformers *ap*-(major) and *sp*-(minor). The crude mixture obtained has been purified by flash column chromatography (hexane: EtOAc = 70:30) in 40% yield and 63% ee. The ee was determined on the non-acetylated product by HPLC analysis on a Daicel Chiralpak AD-H column: hexane/*i*-PrOH 90:10, flow rate 0.9 mL/min, $\lambda = 254$ nm: $\tau_{major} = 23.3$ min, $\tau_{minor} = 37.4$ min. HRMS-ESI-ORBITRAP (+): calculated for $[C_{26}H_{26}NO_5]^+$ 432.1805, found 432.1796 $[M+H]^+$. ^{1}H NMR (600 MHz, $C_2D_2Cl_4$): δ 7.84 (m, 3H), 7.50 (m, 3H), 7.39 (dd, 1H, $J_1 = J_2 = 7.4$ Hz), 7.14 (d, 1H, $J = 8.9$ Hz), 7.06 (d, 1H, $J = 7.7$ Hz), 6.69 (bs, 1H), 6.33 (dd, 1H, $J_1 = 7.7$ Hz, $J_2 = 4.4$ Hz), 3.25 (dd, 1H, $J_1 = 19.4$ Hz, $J_2 = 7.7$ Hz), 3.02 (dd, 1H, $J_1 = 19.4$ Hz, $J_2 = 4.4$ Hz), 1.53 (s, 3H), 1.42 (s, 3H). ^{13}C NMR (150 MHz, $C_2D_2Cl_4$): δ 206.1, 168.2, 158.8, 154.3, 148.5, 135.9, 135.0, 133.4, 132.9, 130.3, 129.4, 128.8, 128.2 (very broad), 127.9, 127.1, 126.0, 125.1, 123.0, 122.9, 80.9, 45.2, 37.2, 28.2, 19.9.

(R)-4-bromo-3-(2-hydroxynaphthalen-1-yl)-2,3-dihydro-1H-inden-1-one (Scheme 4.1—product **222**)

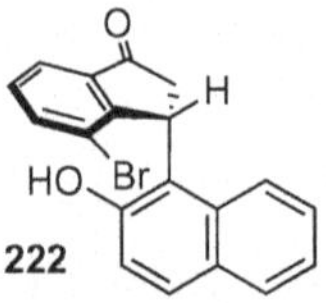

222

The reaction was carried out using catalyst **140**, acid co-catalyst **144** in dry toluene to furnish the crude product **222** as a 86:14 mixture of conformers *ap*-(major) and *sp*-(minor). The crude mixture obtained has been purified by flash column chromatography (hexane:EtOAc = 60:40) in 62% yield and 88% ee. The ee was determined by HPLC analysis on a Daicel Chiralpak AD-H column: hexane/*i*-PrOH 90:10, flow rate 0.75 mL/min, $\lambda = 254$ nm: $\tau_{major} = 14.7$ min, $\tau_{minor} = 19.3$ min. HRMS-ESI-ORBITRAP (+): calculated for $[C_{19}H_{14}BrO_2]^+$ 353.0172, found 353.0180 $[M+H]^+$. ^{1}H NMR (300 MHz, DMSO-d_6): δ 9.46 (s, $1H_A$), 8.42 (d, $J = 8.6$ Hz, $1H_{ap}$),

7.96–7.68 (m, 4H$_{ap}$+xH$_{sp}$), 7.74-7.52 (m, 1H$_{ap}$), 7.46-7.27 (m, 2H$_{ap}$+xH$_{sp}$), 6.99 (d, J = 8.8 Hz, 1H$_{sp}$), 5.77–5.70 (m, 1H$_{sp}$), 5.57–5.46 (m, 1H$_{ap}$), 3.35–3.17 (m, 1H$_{ap}$), 2.67 (dd, J_1 = 18.8 Hz, J_2 = 2.6 Hz, 1H$_{ap}$). ^{13}C NMR (75 MHz, CDCl$_3$): δ 210.7, 188.0, 159.6, 154.2, 142.8, 141.1, 136.7, 132.6, 132.3, 132.1, 130.0, 126.3, 125.9, 125.3, 122.1, 121.3, 47.6, 40.3.

(P)-(R)-1-(7-bromo-3-oxo-2,3-dihydro-1H-inden-1-yl)-8-((tert-butoxycarbonyl)amino)naphthalen-2-yl acetate (Table 4.2—product **224**)

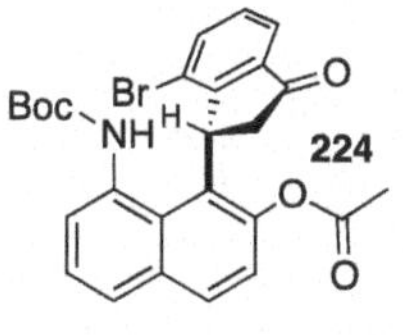

224

The reaction was carried out following the general procedure to furnish the crude product **224** as a single diastereoisomer. The crude mixture obtained has been purified by flash column chromatography (hexane:EtOAc = 70:30) in 55% yield and 88% ee. The ee was determined by HPLC analysis on a Daicel Chiralpak AD-H column: hexane/*i*-PrOH 80:20, flow rate 1.0 mL/min, λ = 254 nm: τ_{major} = 12.0 min, τ_{minor} = 8.8 min. $[\alpha]_{25}^{D}$ = −82.2 (c = 1.0, CHCl$_3$). HRMS-ESI-ORBITRAP (+): calculated for $[C_{26}H_{28}BrN_2O_5]^+$ 527.1176, found 527.1182 $[M+NH_4]^+$. ^{1}H NMR (400 MHz, CDCl$_3$): δ 7.78 (m, 3H), 7.63 (dd, 1H, J_1 = 7.7 Hz, J_2 = 0.9 Hz), 7.55 (bs, 1H), 7.46 (dd, 1H, J_1 = J_2 = 7.7 Hz), 7.27 (dd, 1H, J_1 = J_2 = 7.7 Hz), 7.04 (d, 1H, J = 8.9 Hz), 6.67 (bs, 1H), 6.38 (dd, 1H, J_1 = 8.3 Hz, J_2 = 3.1 Hz), 3.35 (dd, 1H, J_1 = 19.7 Hz, J_2 = 8.7 Hz), 3.06 (bs, 1H), 1.67 (s, 3H), 1.47 (s, 3H). ^{13}C NMR (100 MHz, CDCl$_3$): δ 205.2, 168.2, 156.9, 154.3, 148.4, 139.1, 138.6, 133.0, 130.9, 129.0, 128.7, 128.0 (very broad), 127.0, 125.1, 122.3, 122.2, 121.4, 81.1, 45.1, 38.3, 28.3, 20.4.

(P)-(R)-8-(((benzyloxy)carbonyl)amino)-1-(7-bromo-3-oxo-2,3-dihydro-1H-inden-1-yl)naphthalen-2-yl acetate (Table 4.2—product **230**)

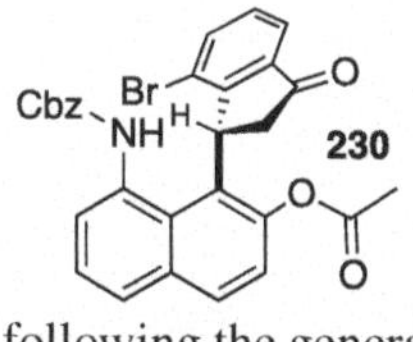

230

The reaction was carried out following the general procedure to furnish the crude product **230** as a single diastereoisomer. The crude mixture obtained has been purified by flash column chromatography (hexane:EtOAc = 2:1) in 41% yield and 82% ee. The ee was determined by HPLC analysis on a Daicel Chiralpak AD-H column: hexane/*i*-PrOH 70:30, flow rate 1.0 mL/min, λ = 254 nm: τ_{major} = 14.6 min, τ_{minor} = 25.2 min. $[\alpha]_{25}^{D}$ = −109.6 (c = 1.0, CHCl$_3$). HRMS-ESI-ORBITRAP (+): calculated for $[C_{29}H_{23}BrNO_5]^+$ 544.0754, found 544.0783 $[M+H]^+$. ^{1}H NMR (300 MHz,

$CDCl_3$): δ 7.78 (m, 3H), 7.67–7.41 (m, 3H), 7.41–7.15 (m, 6H), 7.04 (d, 1H, J = 8.9 Hz), 6.90 (bs, 1H), 6.14 (bs, 1H), 5.19 (bs, 2H), 3.05 (dd, 1H, J_1 = 19.6 Hz, J_2 = 8.3 Hz), 2.88 (bs, 1H), 1.65 (s, 3H). ^{13}C NMR (75 MHz, $CDCl_3$): δ 204.9, 168.2, 156.7, 154.7, 148.5, 139.1, 138.5, 135.9, 133.5, 132.3, 129.7, 129.1, 128.5, 128.3, 128.1, 126.9, 125.1, 122.5, 122.2, 121.4, 67.5, 44.8, 38.3, 20.4.

(P)-(R)-1-(7-bromo-3-oxo-2,3-dihydro-1H-inden-1-yl)-8-iodonaphthalen-2-yl acetate (Table 4.2—product **231**)

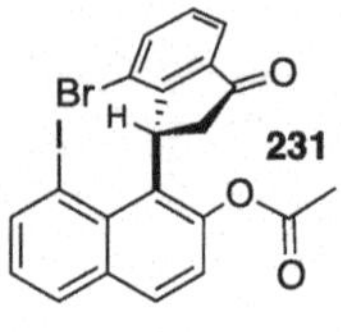

231

The reaction was carried out following the general procedure to furnish the crude product **431** as a single diastereoisomer. The crude mixture obtained has been purified by flash column chromatography (hexane:EtOAc = 80:20) in 40% yield and 97% ee. The ee was determined by HPLC analysis on a Daicel Chiralpak AD-H column: hexane/*i*-PrOH 80:20, flow rate 1.0 mL/min, λ = 254 nm: τ_{major} = 9.9 min, τ_{minor} = 7.9 min. $[\alpha]_{25}^{D}$ = −18.9 (c = 1.0, $CHCl_3$). HRMS-ESI-ORBITRAP (+): calculated for $[C_{21}H_{15}BrIO_3]^+$ 520.9244, found 520.9249 $[M+H]^+$. ^{1}H NMR (300 MHz, $CDCl_3$): δ 8.38 (dd, 1H, J_1 = 7.4 Hz, J_2 = 1.1 Hz), 7.90-7.69 (m, 3H), 7.59 (d, 1H, J = 7.8 Hz), 7.26 (dd, 1H, J_1 = 9.7 Hz, J_2 = 5.6 Hz), 7.10 (m, 2H), 6.52 (dd, 1H, J_1 = 8.8 Hz, J_2 = 3.6 Hz), 3.72 (dd, 1H, J_1 = 19.7 Hz, J_2 = 8.8 Hz), 3.14 (dd, 1H, J_1 = 19.7 Hz, J_2 = 3.6 Hz), 1.74 (s, 3H). ^{13}C NMR (75 MHz, $CDCl_3$): δ 205.1, 168.0, 156.3, 149.0, 143.5, 139.3, 138.7, 134.7, 133.4, 130.3, 130.0, 129.2, 128.0, 126.2, 122.6, 122.0, 121.2, 88.4, 45.2, 39.1, 20.5.

(P)-(R)-8-bromo-1-(7-bromo-3-oxo-2,3-dihydro-1H-inden-1-yl)naphthalen-2-yl acetate (Table 4.2—product **232**)

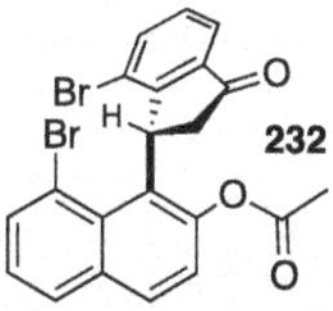

The reaction was carried out following the general procedure to furnish the crude product **232** as a single diastereoisomer. The crude mixture obtained has been purified by flash column chromatography (hexane:EtOAc = 80:20) in 64% yield and 97% ee. The ee was determined by HPLC analysis on a Daicel Chiralpak AD-H column: hexane/*i*-PrOH 95:5, flow rate 0.75 mL/min, λ = 254 nm: τ_{major} = 27.1 min, τ_{minor} = 22.13 min. $[\alpha]_{25}^{D}$ = −26.0 (*c* = 1.0, $CHCl_3$). HRMS-ESI-ORBITRAP (+): calculated for $[C_{21}H_{15}Br_2O_3]^+$ 472.9382, found 472.9380 $[M+H]^+$. ^1H NMR (400 MHz, $CDCl_3$): δ 7.95 (dd, 1H, J_1 = 7.5 Hz, J_2 = 1.2 Hz), 7.86–7.75 (m, 3H), 7.61 (dd, 1H, J_1 = 7.8 Hz, J_2 = 1.0 Hz), 7.33–7.22 (m, 2H), 7.11 (d, 1H, J = 8.9 Hz), 6.62 (dd, 1H, J_1 = 8.8 Hz, J_2 = 3.8 Hz), 3.59 (dd, 1H, J_1 = 19.7 Hz, J_2 = 8.8 Hz), 3.10 (dd, 1H, J_1 = 19.7 Hz, J_2 = 3.8 Hz), 1.69 (s, 3H). ^{13}C NMR (100 MHz, $CDCl_3$): δ 205.1, 168.0, 156.7, 149.0, 139.2, 138.6, 135.0, 134.1, 132.5, 129.8, 129.4, 129.0, 127.9, 125.7, 122.7, 122.1, 121.1, 117.9, 45.1, 39.1, 20.3.

(P)-(R)-1-(7-bromo-3-oxo-2,3-dihydro-1H-inden-1-yl)-8-chloronaphthalen-2-yl acetate (Table 4.2—product **233**)

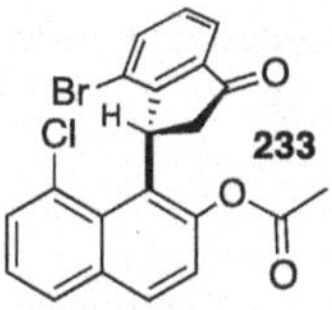

The reaction was carried out following the general procedure to furnish the crude product **233** as a single diastereoisomer. The crude mixture obtained has been purified by flash column chromatography (hexane:EtOAc = 80:20) in 75% yield and 96% ee. The ee was determined by HPLC analysis on a Daicel Chiralpak AD-H column: hexane/*i*-PrOH 95:5, flow rate 0.75 mL/min, λ = 254 nm: τ_{major} = 24.2 min, τ_{minor} = 21.0 min. $[\alpha]_{25}^{D}$ = -15.0 (*c* = 1.0, $CHCl_3$). HRMS-ESI-ORBITRAP (+): calculated for $[C_{21}H_{15}BrClO_3]^+$ 428.9888, found 428.9863 $[M+H]^+$. ^1H NMR (400 MHz, $CDCl_3$): δ 7.79 (m, 3H), 7.69 (dd, 1H, J_1 = 7.6 Hz, J_2 = 1.2 Hz), 7.63 (dd, 1H, J_1 = 7.7 Hz, J_2 = 0.9 Hz), 7.38 (dd, 1H, J_1 = J_2 = 7.7 Hz), 7.27 (dd, 1H, J_1 = J_2 = 7.6 Hz), 7.12 (d, 1H, J = 8.9 Hz), 6.59 (dd, 1H, J_1 = 8.7 Hz, J_2 = 3.9 Hz), 3.52 (dd, 1H, J_1 = 20.1 Hz, J_2 = 8.7 Hz), 3.06 (dd, 1H, J_1 = 20.1 Hz, J_2 = 3.9 Hz), 1.66 (s, 3H). ^{13}C NMR (100 MHz, $CDCl_3$): δ 205.0, 168.1, 156.9, 148.9, 139.1, 138.6,

134.1, 131.3, 130.9, 129.9, 129.7, 129.0, 128.8, 127.6, 125.3, 122.8, 122.2, 121.2, 45.1, 39.1, 20.3.

(P)-(R)-1-(7-bromo-3-oxo-2,3-dihydro-1H-inden-1-yl)-8-((tert-butyldimethylsilyl)oxy)naphthalen-2-yl acetate (Table 4.2—product **234**)

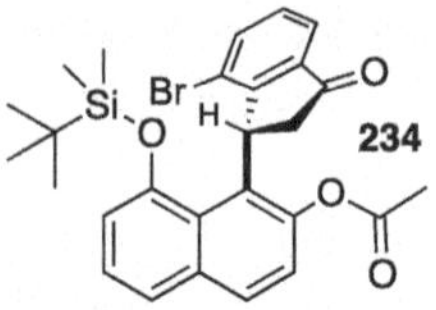

234

The reaction was carried out following the general procedure to furnish the crude product **234** as a single diastereoisomer. The crude mixture obtained has been purified by flash column chromatography (hexane:EtOAc = 85:15) in 82% yield and 94% ee. The ee was determined by HPLC analysis on a Daicel Chiralpak AD-H column: hexane/*i*-PrOH 90:10, flow rate 1.0 mL/min, λ = 254 nm: τ_{major} = 9.8 min, τ_{minor} = 9.1 min. $[\alpha]_{25}^{D} = -161.4$ (c = 1.0, $CHCl_3$). HRMS-ESI-ORBITRAP (+): calculated for $[C_{27}H_{30}BrO_3Si]^+$ 525.1091, found 525.1083 $[M+H]^+$. ^{1}H NMR (300 MHz, $CDCl_3$): δ 7.79 (dd, 1H, J_1 = 7.6 Hz, J_2 = 0.9 Hz), 7.73 (d, 1H, J = 8.9 Hz), 7.66 (dd, 1H, J_1 = 7.6 Hz, J_2 = 0.9 Hz), 7.46 (dd, 1H, J_1 = 8.1 Hz, J_2 = 1.1 Hz), 7.35–7.21 (m, 2H), 7.08–6.96 (m, 3H), 3.30 (dd, 1H, J_1 = 19.7 Hz, J_2 = 8.7 Hz), 2.96 (dd, 1H, J_1 = 19.7 Hz, J_2 = 3.8 Hz), 1.64 (s, 3H), 0.98 (s, 9H), 0.42 (s, 3H), 0.34 (s, 3H). ^{13}C NMR (75 MHz, $CDCl_3$): δ 205.8, 168.4, 157.5, 153.3, 147.4, 138.8, 138.7, 134.5, 129.2, 128.8, 127.8, 126.7, 125.2, 122.5, 122.3, 122.2, 121.6, 115.3, 45.5, 37.9, 26.2, 20.3, 18.9, −3.3, −3.8.

(P)-(R)-1-(7-bromo-3-oxo-2,3-dihydro-1H-inden-1-yl)-8-methoxynaphthalen-2-yl acetate (Table 4.2—product **235**)

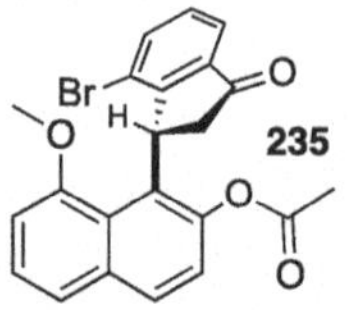

235

The reaction was carried out following the general procedure to furnish the crude product **235** as a single diastereoisomer. The crude mixture obtained has been purified by flash column chromatography (hexane:EtOAc = 75:25) in 93% yield and 90% ee. The ee was determined by HPLC analysis on a Daicel Chiralpak AD-H column: hexane/*i*-PrOH 90:10, flow rate 1.0 mL/min, λ = 254 nm: τ_{major} = 16.7 min, τ_{minor} = 11.9 min. $[\alpha]_{25}^{D} = -66.7$ (c = 0.5, $CHCl_3$). HRMS-ESI-ORBITRAP (+): calculated for $[C_{22}H_{18}BrO_4]^+$ 425.0383, found 425.0354 $[M+H]^+$. ^{1}H NMR (400 MHz, $CDCl_3$): δ 7.80 (dd, 1H, J_1 = 7.7 Hz, J_2 = 1.0 Hz), 7.73 (d, 1H, J = 8.9 Hz), 7.67 (dd, 1H,

$J_1 = 7.7$ Hz, $J_2 = 1.0$ Hz), 7.47 (dd, 1H, $J_1 = 8.3$ Hz, $J_2 = 1.3$ Hz), 7.41 (dd, 1H, $J_1 = J_2 = 8.2$ Hz), 7.27 (ddd, 1H, $J_1 = J_2 = 7.7$ Hz, $J_3 = 0.8$ Hz), 7.05 (d, 1H, $J = 8.9$ Hz), 7.00 (dd, 1H, $J_1 = 7.7$ Hz, $J_2 = 1.2$ Hz), 6.67 (dd, 1H, $J_1 = 8.6$ Hz, $J_2 = 4.1$ Hz), 4.01 (s, 3H), 3.37 (dd, 1H, $J_1 = 19.6$ Hz, $J_2 = 8.6$ Hz), 2.95 (dd, 1H, $J_1 = 19.6$ Hz, $J_2 = 4.1$ Hz), 1.59 (s, 3H). ^{13}C NMR (100 MHz, $CDCl_3$): δ 205.7, 168.5, 158.0, 157.4, 147.2, 138.9, 138.5, 134.0, 128.9, 128.7, 127.8, 125.5, 125.4, 122.4, 122.2, 122.0, 121.6, 107.2, 55.9, 45.5, 39.5, 20.2.

(P)-(R)-1-(7-bromo-3-oxo-2,3-dihydro-1H-inden-1-yl)-8-phenylnaphthalen-2-yl acetate (Table 4.2—product **236**)

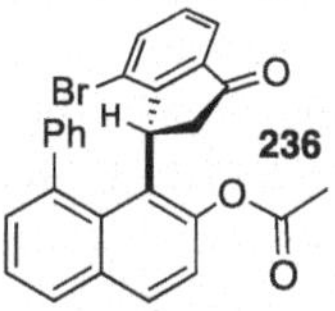

236

The reaction was carried out following the general procedure to furnish the crude product **236** as a single diastereoisomer. The crude mixture obtained has been purified by flash column chromatography (hexane:EtOAc = 80:20) in 58% yield and 94% ee. The ee was determined by HPLC analysis on a Daicel Chiralpak AD-H column: hexane/*i*-PrOH 90:10, flow rate 0.8 mL/min, $\lambda = 254$ nm: $\tau_{major} = 11.4$ min, $\tau_{minor} = 10.7$ min. $[\alpha]_{25}^{D} = -170.3$ ($c = 1.0$, $CHCl_3$). HRMS-ESI-ORBITRAP (+): calculated for $[C_{27}H_{20}BrO_3]^+$ 471.0590, found 471.0573 $[M+H]^+$. ^{1}H NMR (400 MHz, $CDCl_3$): δ 7.86 (m, 2H), 7.71 (m, 1H), 7.62 (dd, 1H, $J_1 = 7.6$ Hz, $J_2 = 0.9$ Hz), 7.53–7.45 (m, 3H), 7.44–7.30 (m, 3H), 7.23–7.12 (m, 2H), 7.07 (d, 1H, $J = 8.6$ Hz), 4.87 (dd, 1H, $J_1 = 8.8$ Hz, $J_2 = 3.2$ Hz), 2.88 (dd, 1H, $J_1 = 19.6$ Hz, $J_2 = 3.2$ Hz), 2.62 (dd, 1H, $J_1 = 19.6$ Hz, $J_2 = 8.8$ Hz), 1.72 (s, 3H). ^{13}C NMR (100 MHz, $CDCl_3$): δ 205.7, 168.3, 156.8, 148.7, 144.8, 139.4, 138.7, 138.5, 133.0, 132.8, 131.5, 129.8, 129.6, 129.3, 129.0, 128.9, 128.3, 128.2, 128.0, 127.2, 124.4, 122.3, 121.7, 121.4, 44.7, 40.2, 20.5.

(P)-(R)-1-(7-bromo-3-oxo-2,3-dihydro-1H-inden-1-yl)-8-(m-tolyl)naphthalen-2-yl acetate (Table 4.2—product **237**)

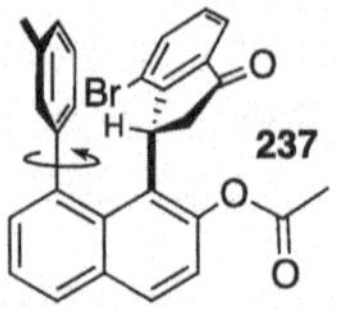

237

The reaction was carried out following the general procedure to furnish the crude product **237** as a mixture 55:45 conformational diastereoisomers due to the slow but not completely blocked rotation of the tolyl group. The crude mixture obtained has been purified by flash column chromatography (hexane:EtOAc = 80:20) in 78% yield and 95% ee. The ee was determined by HPLC analysis on a Daicel Chiralpak AS-H column: hexane/*i*-PrOH 90:10, flow rate 1 mL/min, $\lambda = 254$ nm: $\tau_{major} = 6.8$ min, $\tau_{minor} = 16.9$ min (broad peak). The single peaks obtained for each enantiomer mean that the tolyl group is still rotating, the broadness of the peak means that the time of rotation is comparable to that of the analysis. As additional evidence, we performed the same analysis (on the quasi-racemic compound) at a lower temperature (0 °C) observing broader peaks and at a higher temperature (45 °C) observing narrower peaks. $[\alpha]_{25}^{D} = -112.5$ ($c = 2.0$, $CHCl_3$). HRMS-ESI-ORBITRAP (+): calculated for $[C_{28}H_{22}BrO_3]^+$ 485.0747, found 485.0723 $[M+H]^+$. For the ^{1}H NMR of this compound an integral of “1” has been arbitrarily assigned to the sum of 1 proton from both conformers. ^{1}H NMR (400 MHz, $CDCl_3$): δ 7.85 (m, 2H), 7.62 (m, 1H), 7.49 (m, 3H), 7.39 (m, 1.45H), 7.27 (m, 0.55H), 7.15 (m, 2H), 7.06 (m, 1H), 7.00 (m, 1H), 4.90 (m, 1H), 2.89 (m, 1H), 2.67 (dd, 0.45H, $J_1 = 19.5$ Hz, $J_2 = 8.7$ Hz), 2.53 (dd, 0.55H, $J_1 = 19.5$ Hz, $J_2 = 8.7$ Hz), 2.40 (s, 1.35H), 2.36 (s, 1.65H), 1.73 (m, 3H). ^{13}C NMR (100 MHz, $CDCl_3$): δ 205.7 (double), 168.3 (double), 156.8, 148.7, 148.6, 144.6, 139.6 (double), 139.0, 138.8, 138.7, 138.4 (double), 137.5, 133.0 (double), 132.8, 131.5, 131.3, 130.7, 129.5, 129.3, 128.9 (double), 128.8 (double), 128.3, 128.0, 127.9 (double), 127.0, 125.2, 124.4, 122.2, 121.6, 121.4 (double), 44.8, 44.2, 40.2, 40.1, 21.6, 21.3, 20.5 (double).

(P)-(R)-8-chloro-1-(7-iodo-3-oxo-2,3-dihydro-1H-inden-1-yl)naphthalen-2-yl acetate (Table 4.2—product **238**)

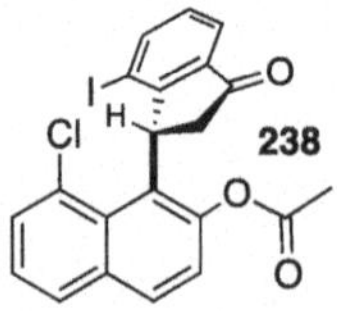

238

The reaction was carried out following the general procedure to furnish the crude product **238** as a single diastereoisomer. The crude mixture obtained has been purified by flash column chromatography (hexane:EtOAc = 80:20) in 35% yield and 93%

ee. The ee was determined by HPLC analysis on a Daicel Chiralpak AD-H column: hexane/*i*-PrOH 80:20, flow rate 1.0 mL/min, $\lambda = 254$ nm: $\tau_{major} = 8.7$ min, $\tau_{minor} = 7.5$ min. $[\alpha]_{25}^{D} = -12.0$ ($c = 1.0$, $CHCl_3$). HRMS-ESI-ORBITRAP (+): calculated for $[C_{21}H_{15}ClIO_3]^+$ 476.9749, found 476.9719 $[M+H]^+$. ^1H NMR (400 MHz, $CDCl_3$): δ 7.91 (dd, 1H, $J_1 = 7.6$ Hz, $J_2 = 1.0$ Hz), 7.82 (m, 3H), 7.71 (dd, 1H, $J_1 = 7.4$ Hz, $J_2 = 1.2$ Hz), 7.39 (dd, 1H, $J_1 = J_2 = 7.8$ Hz), 7.12 (m, 2H), 6.45 (dd, 1H, $J_1 = 8.8$ Hz, $J_2 = 3.8$ Hz), 3.54 (dd, 1H, $J_1 = 19.8$ Hz, $J_2 = 8.8$ Hz), 3.07 (dd, 1H, $J_1 = 19.8$ Hz, $J_2 = 3.8$ Hz), 1.65 (s, 3H). ^{13}C NMR (100 MHz, $CDCl_3$): δ 205.2, 168.0, 160.8, 148.9, 145.2, 138.7, 134.1, 132.3, 131.0, 130.1, 129.9, 129.0, 128.9, 127.6, 125.4, 123.0, 122.9, 94.2, 45.3, 41.7, 20.3.

(P)-(R)-1-(7-iodo-3-oxo-2,3-dihydro-1H-inden-1-yl)-8-(m-tolyl)naphthalen-2-yl acetate (Table 4.2—product **239**)

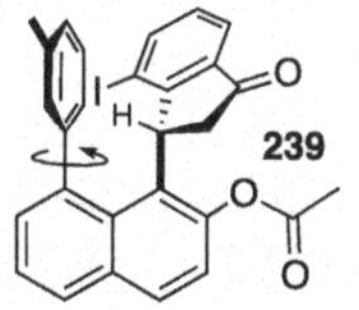

The reaction was carried out following the general procedure to furnish the crude product **239** as a mixture 55:45 conformational diastereoisomers due to the slow rotation of the tolyl substituent. The crude mixture obtained has been purified by flash column chromatography (hexane:EtOAc = 80:20) in 35% yield and 94% ee. The ee was determined by HPLC analysis on a Daicel Chiralpak AS-H column: hexane/*i*-PrOH 80:20, flow rate 1 mL/min, $\lambda = 254$ nm: $\tau_{major} = 5.5$ min, $\tau_{minor} = 10.3$ min (broad peak). $[\alpha]_{25}^{D} = -268.5$ ($c = 1.0$, $CHCl_3$). HRMS-ESI-ORBITRAP (+): calculated for $[C_{28}H_{22}IO_3]^+$ 533.0608, found 533.0602 $[M+H]^+$. For the ^1H NMR of this compound an integral of "1" has been arbitrarily assigned to the sum of 1 proton from both conformers. ^1H NMR (600 MHz, $CDCl_3$): δ 7.86 (m, 2H), 7.79 (m, 1H), 7.66 (m, 1H), 7.56 (m, 1H), 7.48 (m, 1H), 7.42 (m, 1H), 7.38 (m, 0.55H) 7.27 (m, 0.45H), 7.15 (m, 1H), 7.0.7 (m, 1H), 7.00 (m, 2H), 4.77 (m, 1H), 2.89 (m, 1H), 2.68 (dd, 0.45H, $J_1 = 19.8$ Hz, $J_2 = 8.8$ Hz), 2.53 (dd, 0.55H, $J_1 = 19.8$ Hz, $J_2 = 8.8$ Hz), 2.42 (s, 1.35H), 2.37 (s, 1.65H), 1.72 (m, 3H). ^{13}C NMR (150 MHz, $CDCl_3$): δ 205.9 (double), 168.3 (double), 160.8, 160.7, 148.7, 148.6, 145.1 (double), 144.7, 139.8, 139.7, 139.0, 138.4 (double), 137.5, 134.1, 134.0, 132.8 (double), 131.6, 131.4, 130.9, 129.6, 129.3, 129.1, 128.9 (triple), 128.2 (double), 128.0, 127.9 (double), 127.2, 125.4, 124.5 (double), 122.4, 122.3, 94.5, 94.4, 45.0, 44.5, 42.6, 42.5, 21.6, 21.3, 20.5 (double).

(P)-(S)-8-iodo-1-(7-methyl-3-oxo-2,3-dihydro-1H-inden-1-yl)naphthalen-2-yl acetate (Table 4.2—product **240**)

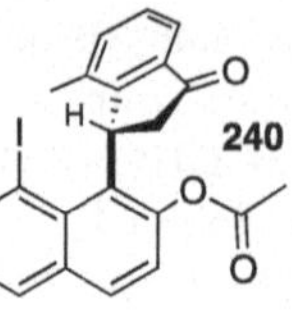

The reaction was carried out following the general procedure to furnish the crude product **240** as a single diastereoisomer. The crude mixture obtained has been purified by flash column chromatography (hexane:EtOAc = 80:20) in 47% yield and 93% ee. The ee was determined by HPLC analysis on a Daicel Chiralpak AD-H column: hexane/*i*-PrOH 80:20, flow rate 1.0 mL/min, λ = 254 nm: τ_{major} = 12.5 min, τ_{minor} = 8.0 min. $[\alpha]_{25}^{D} = -24.1$ (c = 1.0, $CHCl_3$). HRMS-ESI-ORBITRAP (+): calculated for $[C_{22}H_{18}IO_3]^+$ 457.0295, found 457.0286 $[M+H]^+$. ^{1}H NMR (300 MHz, $CDCl_3$): δ 8.39 (dd, 1H, J_1 = 7.4 Hz, J_2 = 1.3 Hz), 7.86 (dd, 1H, J_1 = 8.2 Hz, J_2 = 1.2 Hz), 7.73 (d, 1H, J = 8.9 Hz), 7.67 (d, 1H, J = 7.2 Hz), 7.26 (m, 2H), 7.12 (m, 2H), 6.51 (dd, 1H, J_1 = 8.6 Hz, J_2 = 3.8 Hz), 3.68 (dd, 1H, J_1 = 19.7 Hz, J_2 = 8.6 Hz), 3.10 (dd, 1H, J_1 = 19.7 Hz, J_2 = 3.8 Hz), 1.65 (s, 3H), 1.41 (s, 3H). ^{13}C NMR (75 MHz, $CDCl_3$): δ 206.6, 168.1, 155.9, 149.2, 143.8, 137.1, 136.4, 136.0, 133.6, 130.4, 129.5, 129.0, 127.7, 126.3, 123.0, 120.8, 88.1, 45.4, 37.7, 20.3, 18.1.

(P)-(S)-8-((tert-butoxycarbonyl)amino)-1-(7-methyl-3-oxo-2,3-dihydro-1H-inden-1-yl)naphthalen-2-yl acetate (Table 4.2—product **241**)

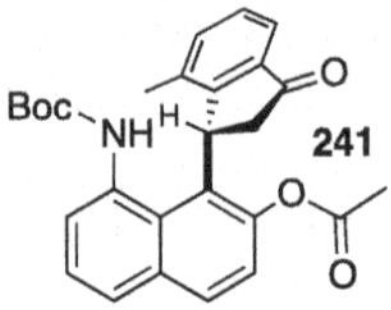

The reaction was carried out following the general procedure to furnish the crude product **241** as a single diastereoisomer. The crude mixture obtained has been purified by flash column chromatography (hexane:EtOAc = 70:30) in 67% yield and 86% ee. The ee was determined by HPLC analysis on a Daicel Chiralpak AD-H column: hexane/*i*-PrOH 80:20, flow rate 1.0 mL/min, λ = 254 nm: τ_{major} = 9.2 min, τ_{minor} = 7.7 min. $[\alpha]_{25}^{D} = -54.4$ (c = 1.0, $CHCl_3$). HRMS-ESI-ORBITRAP (+): calculated for $[C_{27}H_{28}NO_5]^+$ 446.1962, found 446.1920 $[M+H]^+$. ^{1}H NMR (300 MHz, $CDCl_3$): δ 7.76 (m, 2H), 7.62 (m, 2H), 7.46 (dd, 1H, J_1 = J_2 = 7.6 Hz), 7.25 (m, 2H), 7.04 (d, 1H, J = 8.9 Hz), 6.65 (bs, 1H), 6.12 (dd, 1H, J_1 = 8.2 Hz, J_2 = 3.3 Hz), 3.35 (dd, 1H, J_1 = 19.3 Hz, J_2 = 8.2 Hz), 3.07 (dd, 1H, J_1 = 19.3 Hz, J_2 = 3.3 Hz), 1.59 (s, 3H), 1.45 (s, 9H). ^{13}C NMR (75 MHz, $CDCl_3$): δ 206.5, 168.3, 156.7, 154.1, 148.6, 136.7, 136.4, 136.2, 133.5, 133.0, 130.0, 129.1, 128.4 (broad), 127.8, 127.5 (broad), 125.3, 122.5, 120.8, 81.1, 45.0, 37.4, 28.3, 20.2, 18.0.

(P)-(S)-8-(((benzyloxy)carbonyl)amino)-1-(7-methyl-3-oxo-2,3-dihydro-1H-inden-1-yl)naphthalen-2-yl acetate (Table 4.2—product **242**)

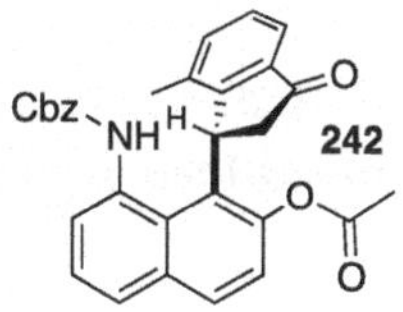

242

The reaction was carried out following the general procedure to furnish the crude product **242** as a single diastereoisomer. The crude mixture obtained has been purified by flash column chromatography (hexane:EtOAc = 60:40) in 56% yield and 83% ee. The ee was determined by HPLC analysis on a Daicel Chiralpak AD-H column: hexane/*i*-PrOH 60:40, flow rate 0.75 mL/min, $\lambda = 254$ nm: $\tau_{major} = 15.4$ min, $\tau_{minor} = 20.8$ min. $[\alpha]^{D}_{25} = -46.2$ ($c = 1.0$, $CHCl_3$). HRMS-ESI-ORBITRAP (+): calculated for $[C_{30}H_{26}NO_5]^+$ 480.1805, found 480.1794 $[M+H]^+$. ^1H NMR (400 MHz, $CDCl_3$): δ 7.76 (m, 2H), 7.60 (m, 2H), 7.46 (dd, 1H, $J_1 = J_2 = 7.8$ Hz), 7.30–7.10 (m, 7H), 7.03 (d, 1H, $J = 8.9$ Hz), 6.82 (bs, 1H), 5.90 (bs, 1H), 5.26–5.08 (m, 2H), 3.16 (dd, 1H, $J_1 = 18.9$ Hz, $J_2 = 8.7$ Hz), 2.94 (dd, 1H, $J_1 = 18.9$ Hz, $J_2 = 3.8$ Hz), 1.56 (s, 3H), 1.41 (s, 3H). ^{13}C NMR (100 MHz, $CDCl_3$): δ 206.1, 168.2, 156.5, 154.8, 148.8, 136.9, 136.4, 136.3, 135.9, 133.6, 132.4, 130.1, 129.1, 128.8, 128.4, 128.3, 128.1, 127.8, 127.6, 125.3, 122.7, 120.8, 67.5, 44.8, 37.5, 20.1, 17.9.

(P)-(S)-8-iodo-1-(3-oxo-7-phenyl-2,3-dihydro-1H-inden-1-yl)naphthalen-2-yl acetate (Table 4.2—product **243**)

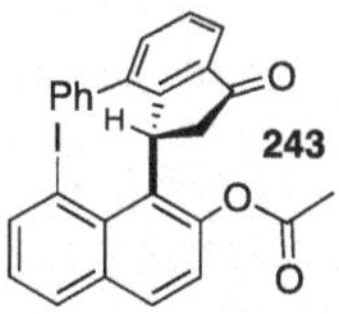

243

The reaction was carried out following the general procedure to furnish the crude product **243** as a single diastereoisomer. The crude mixture obtained has been purified by flash column chromatography (hexane:EtOAc = 80:20) in 15% yield and 98% ee. The ee was determined by HPLC analysis on a Daicel Chiralpak AD-H column: hexane/*i*-PrOH 80:20, flow rate 1.0 mL/min, $\lambda = 254$ nm: $\tau_{major} = 8.4$ min, $\tau_{minor} = 10.6$ min. $[\alpha]^{D}_{25} = -29.4$ ($c = 0.5$, $CHCl_3$). HRMS-ESI-ORBITRAP (+): calculated for $[C_{27}H_{20}IO_3]^+$ 519.0452, found 519.0431 $[M+H]^+$. ^1H NMR (300 MHz, $CDCl_3$): δ 8.15 (dd, 1H, $J_1 = 7.3$ Hz, $J_2 = 1.2$ Hz), 7.81 (dd, 1H, $J_1 = 7.6$ Hz, $J_2 = 1.1$ Hz), 7.48 (dd, 1H, $J_1 = 8.2$ Hz, $J_2 = 1.2$ Hz), 7.41 (dd, 1H, $J_1 = J_2 = 7.5$ Hz), 7.36–7.29 (m, 2H), 6.93 (m, 2H), 6.78 (m, 2H), 6.66–6.54 (m, 4H), 3.69 (dd, 1H, $J_1 = 19.5$ Hz, $J_2 = 8.8$ Hz), 3.15 (dd, 1H, $J_1 = 19.5$ Hz, $J_2 = 3.7$ Hz), 1.75 (s, 3H). ^{13}C NMR (75 MHz, $CDCl_3$): δ 206.7, 167.8, 156.1, 149.3, 142.4, 140.6, 138.1, 137.5, 135.8, 133.9, 132.9, 129.6, 129.2, 128.9, 127.8, 127.4, 127.0, 125.9, 125.6, 122.2, 121.9, 88.5, 44.6, 37.3, 20.6.

(P,M)-(R)-1-(7-bromo-3-oxo-2,3-dihydro-1H-inden-1-yl)-8-(phenanthren-9-yl)naphthalen-2-yl acetate and **(P,P)-(R)-1-(7-bromo-3-oxo-2,3-dihydro-1H-inden-1-yl)-8-(phenanthren-9-yl)naphthalen-2-yl acetate** (Table 4.2—product **244a** and **244b**)

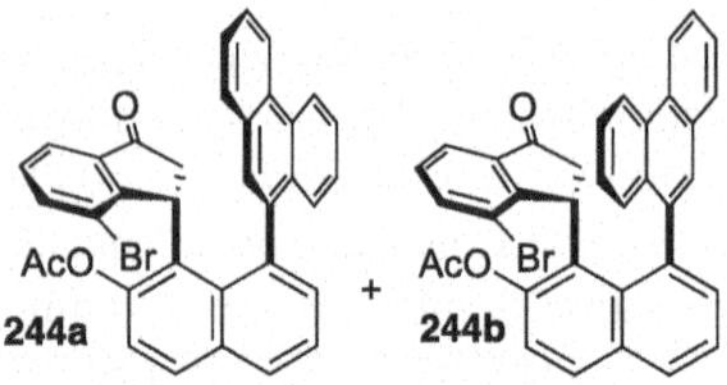

The reaction was carried out following the general procedure to furnish the crude products ***(R,P,M)*-244a** and ***(R,P,P)*-244b** as a mixture 50:50 diastereoisomers due to the completely blocked rotation of the phenantryl group. The crude mixture obtained has been purified by flash column chromatography (hexane:EtOAc = 80:20) in 15% yield and 95% and 65% ee. The ee was determined by HPLC analysis on a Daicel Chiralpak AD-H column: hexane/*i*-PrOH 90:10, flow rate 1 mL/min, ***(R,P,M)*-244a** $\lambda = 254$ nm: $\tau_{major} = 12.0$ min, $\tau_{minor} = 32.6$ min; ***(R,P,M)*-244b** $\lambda = 254$ nm: $\tau_{major} = 20.2$ min, $\tau_{minor} = 13.6$ min. $[\alpha]^{D}_{25} = -341.0$ ($c = 1.0$, $CHCl_3$, mixture of diastereoisomoers). HRMS-ESI-ORBITRAP (+): calculated for $[C_{35}H_{24}BrO_3]^+$ 571.0903, found 571.0904 $[M+H]^+$. For the ^{1}H NMR of this compound an integral of "1" has been arbitrarily assigned to the sum of 1 proton from both conformers. ^{1}H NMR (300 MHz, $CDCl_3$): δ 8.80–8.65 (m, 2H), 8.31 (dd, 0.5H, $J_1 = 8.1$ Hz, $J_2 = 1.4$ Hz), 8.10–7.88 (m, 3H), 7.80 (dd, 0.5H, $J_1 = 7.5$ Hz, $J_2 = 1.9$ Hz), 7.72–7.32 (m, 9H), 7.20–7.33 (m, 2H), 5.09 (dd, 0.5H, $J_1 = 8.7$ Hz, $J_2 = 3.0$ Hz), 4.73 (dd, 0.5H, $J_1 = 8.8$ Hz, $J_2 = 3.4$ Hz), 2.73 (dd, 0.5H, $J_1 = 19.6$ Hz, $J_2 = 3.3$ Hz), 2.10 (m, 1H), 1.63 (s, 3H), 1.06 (dd, 0.5H, $J_1 = 19.9$ Hz, $J_2 = 8.8$ Hz). ^{13}C NMR (75 MHz, $CDCl_3$): δ 205.9, 205.3, 168.3 (double), 156.7, 156.4, 148.5, 141.1, 139.5, 138.8 (double), 138.5, 138.3, 137.4, 136.9, 134.6, 133.6, 133.2, 132.6, 131.8, 131.6, 131.0, 130.1, 130.0, 129.9, 129.8, 129.7, 129.1, 128.9, 128.8, 128.7, 128.6, 128.5, 128.3, 128.2, 127.6, 127.5, 127.3, 127.2, 127.1, 126.9, 126.8, 125.2, 124.3, 123.1, 122.9, 122.7, 122.6, 122.5, 122.3, 121.6, 121.5, 121.4, 121.1, 44.2, 43.8, 41.2, 38.9, 20.5, 20.4.

(P)-8-bromo-1-((1R,3R)-7-bromo-3-hydroxy-2,3-dihydro-1H-inden-1-yl)naphthalen-2-ol (Reaction 4.3—product **245**)

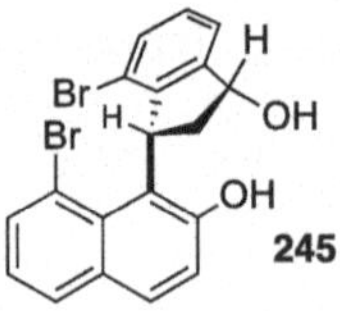

245

A large scale reaction was developed following the general procedure to give, after column chromatography on silica gel, compound **232** in 83.3% yield (1.71 mmol). Compound **232** (1.71 mmol, 1 equiv.) was then dissolved in MeOH (115 mL, 0.015 M) and $NaBH_4$ (640 mg, 17.1 mmol, 10 equiv.) was added. The resulting mixture was heated to 60 °C and left stirring overnight. The next day the reaction was quenched with ice-cold water and extracted with DCM. The organic fractions were made anhydrous over $MgSO_4$, concentrated under vacuum and purified with flash column chromatography (hexane:EtOAc 70:30) to obtain the title compound in 87.7% yield. HRMS-ESI-ORBITRAP (+): calculated for $[C_{19}H_{15}Br_2O_2]^+$ 432.9433, found 432.9428 $[M+H]^+$. ^{1}H NMR (600 MHz, DMSO-d_6): δ 9.69 (bs, 1H), 7.83 (m, 2H), 7.75 (d, 1H, $J = 9.2$ Hz), 7.37 (d, 1H, $J = 7.7$ Hz), 7.20 (d, 1H, $J = 7.7$ Hz), 7.16-7.01 (m, 3H), 5.97 (m, 1H), 5.59 (bs, 1H), 5.16 (m, 1H), 3.02 (m, 1H), 2.48 (m, 1H, partially overlapped with DMSO). ^{13}C NMR (100 MHz, DMSO-d_6): δ 156.3, 148.8, 144.9, 134.8, 132.8, 131.4, 131.3, 130.2, 129.9, 128.1, 123.7, 123.1, 119.7, 119.5, 117.6, 116.8, 73.3, 42.8, 42.7.

References

1. Di Iorio N, Filippini G, Mazzanti A, Righi P, Bencivenni G (2017) Controlling the C(sp^3)–C(sp^2) axial conformation in the enantioselective Friedel-Crafts-type alkylation of β-Naphthols with inden-1-ones. Org Lett 19(24):6692–6695. https://doi.org/10.1021/acs.orglett.7b03415
2. Gustafson JL, Lim D, Miller SJ (2010) Science, 1251
3. Link A, Sparr C (2014) Angew Chem Int Ed, 5458
4. Staniland S, Adams RW, Grainger JJW, Turner NJ, Clayden J (2016) Angew Chem Int Ed, 10755
5. Joliffe JD, Armstrong RJ, Smith MD (2017) Nat Chem. https://doi.org/10.1038/nchem.2710. (article in press)
6. Narute S, Parnes R, Toste DF, Pappo D (2016) J Am Chem Soc, 16533
7. Li GQ, Gao H, Keene G, Devonas M, Ess DH, Kurti L (2013) J Am Chem Soc, 7414
8. Chen YH, Cheng DJ, Zhang J, Wang Y, Liu XY, Tan B (2015) J Am Chem Soc, 15062
9. Zhang H–H, Wang C–S, Li C, Mei G–J, Li Y, Shi F (2016) Angew Chem Int Ed, 116
10. Moliterno M, Cari R, Puglisi A, Antenucci A, Sperandio C, Moretti E, Di Sabato A, Salvio R, Bella M (2016) Angew Chem Int Ed, 6525
11. Brandes S, Bella M, Kjoersgaard A, Jørgensen KA (2006) Angew Chem Int Ed, 1147
12. Lomas JS, Dubois J–E (1976) J Org Chem, 3033
13. Lomas JS, Luong PK, Dubois J–E (1977) J Org Chem, 3394
14. Lomas JS, Anderson JE (1995) J Org Chem, 3246
15. Wolf C, Pranatharthiharan L,. Ramagosa RB (2002) Tetrahedron Lett, 8563
16. Casarini D, Lunazzi L, Mazzanti A (1997) J Org Chem, 3315
17. Ford TW, Thompson TB, Snoble KAJ, Timko JM (1975) J Am Chem Soc, 95
18. Paradisi PR, Mazzanti A, Ranieri S, Bencivenni G (2012) Chem Commun, 11178
19. Still WC, Kahn M, Mitra AJ (1978) J Org Chem 43:2923

Chapter 5
Conclusions and Future Outlooks

It is somehow difficult to give a general conclusion to a thesis made of single projects that are different between each other. To do that, we must keep in mind the *leitmotiv* of the entire work: organocatalysis.

As stated in the preface, we wanted to develop new reactivities and achieve specific structural features in a molecule by means of asymmetric organocatalysis and we can certainly conclude that we succeded in our purpose. This elaborate added yet another small piece to this vast and promising branch of chemistry. We contributed with new structures and reactions in a clean, practical and efficient way, in agreement with the principles of organocatalysis.

In spite of the good results we obtained, there are still countless applications for our products and many possible outlooks in the field of organic catalysis. For instance, further experiments are in progress in our own laboratory for the use of the atropisomeric diol **245** as a chiral ligand together with other projects, all exploiting the unlimited possibilities that organocatalysis offers.

N. Di Iorio, *New Organocatalytic Strategies for the Selective Synthesis of Centrally and Axially Chiral Molecules*, Springer Theses,
https://doi.org/10.1007/978-3-319-74914-3_5

The manufacturer's authorised representative in the EU is Springer Nature Customer Service Centre GmbH, Europaplatz 3, 69115 Heidelberg, Germany. If you have any concerns regarding our products, please contact ProductSafety@springernature.com

Printed and bound by CPI Group (UK) Ltd, Croydon, CR0 4YY
05/10/2026
02228139-0020